SHY BOY

SHY BOY

The Horse That Came In from the Wild

MONTY ROBERTS

WITH PHOTOGRAPHS BY
CHRISTOPHER DYDYK

Perennial

An Imprint of HarperCollins*Publishers*

A hardcover edition of this book was published in 1999 by HarperCollins Publishers.

SHY BOY. Text and photographs copyright © 1999 by Monty Roberts. All photographs by Christopher Dydyk, with the following exceptions: Blushing ET on page 176–77 copyright © Benoit Photo; photos on pages 6–7, 10–11, 16–17, 20, 24–25, 29, 35 copyright © Gary Leppart.

HarperCollins books may be purchased for educational, business, or sales promotional use. For information please write: Special Markets Department, HarperCollins Publishers, Inc., 10 East 53rd Street, New York, NY 10022.

First Perennial edition published 2000.

Designed by Elliott Beard

ISBN 0-06-093289-9

00 01 02 03 ❖/RRD 10 9 8 7 6 5 4 3 2

Being around horses poses inherent dangers. Working with wild horses is generally more dangerous than working with domestic horses, so every precaution should be taken to be as safe as possible, should you decide to act hands-on. Seek professional help, learn the language of Equus, and exercise the greatest of caution any time you're on or around horses.

*I dedicate this book to the American mustang,
the heart and soul of my work. Shy Boy represents
the wild horse in a truly honorable fashion.*

ACKNOWLEDGMENTS

Many people have assisted me in the creation of this book. Christopher Dydyk is a talented young man who produces photographs that are indeed works of art. If he continues to work as hard as he has in the past, his talents will be widely recognized the world over.

My wife, Pat, set aside her own creative endeavor—equine sculpture—for many weeks, to help with the organization of words and photographs to bring this project to completion. Our daughter, Laurel, assisted with the Shy Boy adventures and was most helpful in bringing together the necessary equipment and people to get the job done. Our son, Marty, has set aside his legal career to run our family operation and it is through his efforts that all of this is possible. To these family members I owe the biggest thanks possible.

A team including many others were most helpful and should be acknowledged. Susan Watt in London and Trena Keating in New York have represented HarperCollins Publishers. Trena Keating, it should be noted, took a hands-on approach and was most active in helping to sculpt the American version. She took the time to come to see events, to meet Shy Boy, and to get a feel for the book in a most personal way.

Louise Dennys, Knopf Canada, continues to be the author's dream. Consistently, Louise steps up with solutions. Louise's friend and fellow Canadian, Lawrence Scanlan, who so successfully edited my first book, again offered advice with exceptional insight and sensitivity.

My literary agent, Jane Turnbull, is an ever present source of support, and the acceptance by the publishers of my first two books certainly would not have been the same without her efforts.

Caroline Baldock, my literary assistant, continues to help Pat and myself as I work to get my concepts immortalized on pages. Whether Caroline is in California, London, or Germany, she works very long days to get my words on paper in the best way possible.

Teams of people on each of the projects referred to in this book gave great effort in so many ways, the absence of which would have significantly reduced the enjoyment of the book. To these and all the aforementioned people I send the strongest debt of gratitude I can muster from within me.

PROLOGUE

It seemed we were following a ghost . . .

The story of the mustang I would come to call Shy Boy goes back several years, though he himself could not have known that. For the first years of his life he was concerned only with the politics of the herd and the sometimes stark business of survival. The world for this wild horse began and ended with the high desert beyond the Sierra Nevada range, with its wide, majestic horizons.

The little bay horse was born, I know, somewhere in the hills beyond Tonopah, Nevada. Geographical place names nearby reveal something of the territory—Lone Peak, Cactus Mountain, Mud Lake, Willow Creek, the Silver Peak Range. Shy Boy and the mustangs in his herd would have known them only as places to find water, shelter from winter winds, valleys with decent grazing in dry summers.

I had had my eye on Shy Boy for forty-five years. Not on him, exactly, but on one of his kind. He was a vision, but it would take good fortune and hard work to bring us together.

My first book—*The Man Who Listens to Horses*—recounted

my life, the sum total of my experience, and the reasoning behind the message I am always striving to deliver: let us make the world a better place for the horse—all horses, including free horses without names.

One such horse became Shy Boy to me on the night of March 30, Easter Sunday, 1997. I was in the high country of central California attempting to make a point: that through communication and fair treatment, the wild horse will forge a relationship with people. I wanted this little mustang to show the world—because the movie cameras were rolling—that when treated with respect he would show his capacity for generosity and trust.

A lot was on the line that night, for I had another reason for being there: as a teenage boy, I had gentled wild horses alone in the very landscape that Shy Boy would have recognized as home. I wanted to prove something to the doubting Thomases, both latter-day and modern. I had gentled wild horses once up there; I would do it again.

That night in the Cuyama Valley, the moon lit our private world. Just two horses—one domestic, one wild—and me.

Big Red Fox, my tall bay horse, a rugged nine-year-old Thoroughbred retired from the racetrack, settled into a routine. When the mustang moved, he moved. When the wild horse stopped to graze or take water, Big Red Fox also stopped. Foxy quickly learned this game and I soon saw no need to direct him. He was maintaining a position some fifty yards from the mustang.

Around ten P.M. I used my walkie-talkie to call Caleb Twisselman, a gifted young rider I had enlisted to help me, and asked him to bring me some more clothes. It was getting cold. Directly overhead Comet Hale-Bopp was dragging its enormous tail across the night sky. In that thin air, with no city lights to dilute the darkness, it seemed you might reach out and pluck that comet from the sky. I watched it proceed westward, then I lowered my line of sight and saw the mist.

A marine layer was creeping up from the Pacific Coast, maybe fifty miles away and 3,500 feet below the valley. The mist was like the sea itself, rising up to envelop us.

At eleven P.M. the mist took away the light of the moon, as if someone had flipped a switch. Owing to what some might call a flaw in my vision, I see no color, only shades of gray. It means I see better than most people in the dark, yet I could not see a thing. Suddenly the natural light we had planned on was lost and we were staring disaster in the face.

I was afraid we were going to lose this mustang, that he would come to a fence and find a way through or over, and that we would spend the whole week searching for him. But that night the extraordinary skill of my saddle horse let me stay in touch with his wild cousin.

Big Red Fox kept me on the mustang's path; he would stop, start, take a right, hang a left, with no instruction from me. I could have been riding with my eyes shut. It was as though the mustang was wearing a tracking device. Big Red Fox could see the mustang, smell him or sense him—I do not have the answer to this miracle. However he did it, he kept on him because every now and then I would hear the click of a hoof on stone or see a faint silhouette in the mist.

It raised the hairs on the back of my neck, this eerie game of tag. I would see a shadow, Big Red Fox would take me through it, and it would turn out to have been the mustang. It seemed we were following a ghost. For Big Red Fox, at least, the trail was luminous and clear.

During these shadowy glimpses of the mustang, it looked to me that Shy Boy was crouching, keeping his head and neck low, using every ravine and crevice to make himself harder to see. I found myself repeating the words, "Hey, Little Shy Boy, are you trying to get lost from me?"

I was shivering with cold and this phrase turned into a song that I said aloud over and over. The song led to the mustang's name, for it went, in part,

Shy Boy, Shy Boy, don't be so shy
I know we'll be friends in the by and by.

I uttered those words partly to keep my energy focused, partly to take my mind off freezing, mostly to let the wild horse know from the kindness in my voice that I meant him no harm.

SHY BOY

1

SHY BOY'S TRIBE

*Wild horses have become living symbols—of liberty
and beauty and power.*

To appreciate Shy Boy, you must see him or his kind running, free and easy, in a wide-open space. To see what I mean, look at the photograph on pages 198–99: "Shy Boy enjoying a wild gallop on a spring morning."

A free horse running is a beautiful sight. The long tail high and proud. The mane rising and falling with his rolling step. Domestic horses striding in paddocks, without saddles and riders, are a pleasure to behold. But a mustang herd running through sagebrush, running just to feel the wind: that sight resonates in us all. I have seen it countless times and it always stirs my heart.

The estimated 30,000 wild horses that now roam reserves are remnants of herds on the Great Plains that once numbered in the millions.

The magnificent herds of wild horses that once roamed the North American Plains in the millions no longer exist. Mustangs are somewhat rare now, and for that reason, cherished. Shy Boy and his tribe have become living symbols—of liberty and beauty and power.

To some extent we attach to the mustang the same characteristics we would like to see in ourselves: strong, wild at heart, sensitive, graceful—and above all, free. All over the world, people view the wild horse as a noble and romantic animal; in America, especially, the mustang is an icon.

The Lone Ranger's famous white horse, Silver, we were led to believe, was a mustang stallion wounded by a buffalo and then nursed back to health by the ranger. Old western films had cowboys coming to the rescue of wild horses, who paid them back by accepting the saddle and riding like the wind.

One automobile manufacturer even named a car after the mustang. The chrome silhouette of the galloping horse—not unlike the image of Shy Boy on pages 198–99—was set in the car's front grill, and the car was pitched to a generation as small, fast, and youthful.

The odd thing is this: the romantic attachment to mustangs existed even in the minds of pioneers in the nineteenth century, when wild horses on the Plains were as common as sparrows. Matt Field, a traveler on the Santa Fe Trail in 1839, wrote after admiring one striking sorrel stallion that "a domestic horse will ever lack that magic and indescribable charm that beams like a halo around the simple name of freedom. . . . He was free, and we loved him for the very possession of that liberty we longed to take from him."

After the West had been settled, cattle arrived, fences went up, and the rich grasslands where the mustang roamed were no longer theirs. The herds sought refuge in higher, tougher ground. Ever since then there has been a pitched battle between those who want available land put to practical use (ranching and hunting) and those who want land set aside for the ever-diminishing herds of wild horses.

Shy Boy's ancestors have long been on the losing side of that old turf war. More than one million wild horses were captured by the government for use in World War I; hundreds of thousands more were taken to abattoirs and used in animal feeds; some were shot for sport. In one especially brutal killing in December 1998, thirty-four free-ranging horses in the foothills of the Virginia Range in Nevada were shot; it was clear from the aftermath that whoever did it meant for the horses to suffer and to die slowly. The atrocity made headlines around the world. In the West there was shock and outrage—on both sides of the mustang issue. A sacred line had been crossed. "It's like somebody desecrating the flag," one Nevada investigator said.

The mustang has become a kind of conscience of America. What we've done to horses has suddenly become a factor in our lives: the strong link between cruelty to animals and cruelty to people has been well documented.

Pages 6–7: **The land left to mustangs is often rugged high country where severe winters and poor grazing sometimes take their toll on even the hardy mustang.**

It should not surprise that people feel the way they do about horses. No other animal in history has had more impact on our lives than the horse. Millions of horses gave their lives in our wars. They transported settlers in covered wagons across continents, delivered our mail, plowed our fields, cleared our lands, and entertained us with their athleticism on racetracks and show grounds. And during all this time they have provided faithful companionship.

Shy Boy's ancestors were brought to North America by Spanish conquistadors four centuries ago. In a way, this marked a homecoming for the horse. Horses and their ancestors had developed on this continent 55 million years before; from here they crossed land bridges into Asia and spread to Europe. By the time Columbus landed in North America, the indigenous horse had been extinct for almost fifteen thousand years. No one is really sure why they disappeared. Meanwhile, slowly but surely, Shy Boy's ancestors had circumnavigated the globe.

Horses abandoned or lost by Spanish cavalry, the troops of the conquistadors, and later by Spanish settlers were the forebears of the mustangs that colonized the wild heartland of the western United States. They were tough horses; they had to be to survive the grueling eight-week journey by ship to the Americas from Spain. Travel in those days was not for the fainthearted, animal or human.

Some believe that the "horse latitudes" (30 degrees north latitude—the Tropic of Cancer; 30 degrees south latitude—the Tropic of Capricorn), zones where ships were often becalmed, owe their names to this fact: when water ran out and horses on board died of thirst, they were tossed into the sea. The difficul-

ties in transporting livestock across the ocean led the Spanish to establish horse-breeding farms in Cuba.

Once on land, the horses went to work—as warhorses. Hernando Cortez was recorded as saying that next to God, he owed victory to the horse. Horse and soldier climbed mountains, forded rivers and swamps, battled through impenetrable undergrowth, and fought indigenous peoples every step of the way. If the men were made of iron, the horses were forged of something even stronger.

Bernal Díaz del Castillo, who accompanied Cortez in 1519, recorded the colors of the expedition's horses: dark and light chestnuts, light bays, dappled, almost black, piebald, golden bay, perfect bay, mahogany bay, brown, and black.

Exactly how and when the horse reinhabited the Great Plains and the high desert areas of the North American continent has been the focus of great debate. Free-ranging horses are recorded as early as 1841 in the Great Basin area in Idaho. The explorer John Charles Fremont recorded in his dairy in 1843 that horse tracks had been found along the edge of Pyramid Lake in Nevada. Certainly, Native Americans knew enough to steal the horse belonging to Kit Carson, who was a member of Fremont's expedition.

Meanwhile, in the 1800s wild horses were rounded up for ranch duty. Cowboys learned to work horses and cattle from the Spanish, who had, in turn, learned their exquisite horsemanship from the Moors when they conquered Spain. In the

Pages 10–11: **Wild horses at sunset. The sight of free horses running has the power still to stir our hearts.**

American West, cowboys adapted the Spanish riding style to their needs, bringing the leg position farther forward and changing the cut of the Spanish saddle. The aim was to sit the rider further forward on the horse, over the horse's powerful shoulders. The result was a more athletic horse, especially important in rough or mountainous terrain.

Cowboys also had to develop a way of guiding the horse that left a hand free for working a rope. This is why they learned to neck-rein the horse, quite a departure from the English riding style, in which the rider guides the horse by a direct rein in each hand linked to the horse's mouth. The neck rein is essentially a loose rein; to turn a horse left, you simply apply the rein to the right side of the horse's neck. Cowboys held the reins above the pommel, leaving the other hand free to work the rope. To get this free hand, they had to develop a control system that was quick, accurate, and, of course, one-handed.

Ranch owners in California employed Spanish horsemen, called "reinsmen," to train horses to work cattle. Horses trained by reinsmen could do extraordinary things: they could stop in an instant from a full gallop; they could turn a steer as it was running along a fence; they could cut an individual cow out from the herd and hold it out, bringing it under control. Reinsmen used bits that were heavier and more complicated than the simple snaffle (a very gentle and basic bit), and reinsmen trained their horses to give an immediate, dynamic response from the lightest signal.

The word "mustang" is derived from the Spanish word *mestengo,* or "stray beast." These early stray animals quickly reverted to a wild state. An amazing aspect of horses is how quickly a domesticated horse can revert to its purest, natural

instincts and become as wild as its ancestors were thousands of years ago. And yet a gentled mustang, in the right hands, can just as quickly become a capable and worthy mount.

Had governments not passed laws protecting the wild horse, Shy Boy's tribe might have disappeared entirely. In 1971, the United States Congress passed a law to protect wild horses and burros. There are an estimated 30,000 free-roaming mustangs in the western United States; Canada has isolated pockets of wild horses but their numbers are thought to be low.

The U.S. government set about to charge various agencies with maintaining appropriate numbers of mustangs on federal lands. This responsibility eventually was assigned to the federal Bureau of Land Management (BLM). Eighty million acres in sixteen states were placed under strict management to protect these wild animals. Before the law was passed, more letters poured into Congress over the plight of wild horses than had been generated by any other issue in U.S. history. One congressman is said to have received fourteen thousand letters in the days leading up to the final vote. A measure of where mustangs ranked in American hearts and minds is that the law passed without a single dissenting vote.

Since my first contact with the American mustang in the late 1940s, I have schooled hundreds of them, both in open country and in enclosures. I have as much respect for the mustang as for any other breed I have worked with, and that includes Thoroughbreds, quarter horses, Arabians, warm-bloods (a Thoroughbred crossed with a heavy horse), most of the pony breeds, and several draft breeds—the big farm horses known for their pulling power.

The American mustang has become a diversified genetic

entity: tough, highly intelligent, and blessed with physical attributes often superior to most of our domestic breeds. Mustangs' physiological makeup tends to help them overcome illness or injury; in this regard, the domestic horse pales in comparison. But for all his genetic diversity, the mustang's unique gait at trot and canter still points to his Spanish and North African ancestry. Like the barbs of North Africa, the running mustang tends to lift his feet high and flair them to the outside in a kind of paddling motion.

Most mustangs are astonishingly hardy; they've had to be to survive in the wild. The endurance capabilities of American mustangs are legendary. They can go all day for you, and day after day as well, with a generosity seldom seen in the domestic horse.

One cannot list the attributes of the American mustang without making special note of his feet. Just as walking barefoot encourages tough callused feet in humans, the mustang owes the thick walls and soles of his feet to environmental factors. Day after day of walking and running over rock and hard ground produce remarkably sturdy feet, feet that were already strong at birth. Breeds such as the barbs of North Africa and the Andalusians of Spain, the mustang's genetic ancestors, are also noted for their fine and durable feet.

I believe that the wild American mustang should be preserved as a national treasure. Their numbers should be watched over to allow them to live along with other species, utilizing the range lands where they currently exist. I find no better way to see to their best interests than through the existing Bureau of Land Management, the agency that protects America's wild mustang herds.

Not everyone agrees on the correct policy toward America's

wild horses. There are those who would kill all of the mustangs; others would encourage them to roam free without controlling their numbers. If mustang numbers dip too low, they risk disappearing; on the other hand, there is only so much land to provide appropriate territory for wild horses and if their numbers accelerate the result is starving horses. It seems to me that the best course lies somewhere between these two positions. The BLM runs an adoption program, whereby individual citizens can acquire wild horses. The aim is twofold: to help control mustang numbers in the wild and to put mustangs in the hands of people who will treat them well.

I support the BLM's adoption program in principle, and I have a great deal of sympathy for the BLM itself, but the agency should do more to gentle horses before they're adopted. People are often ill-prepared to deal with a wild horse at home. The adoption program, then, needs significant revision.

The BLM has not always made the right decisions—choosing which horses to put up for adoption and which to return to the wild, for example, is not always well done. But there is only so much one agency can do, and many of the BLM's decisions are good ones. Some small groups—though well-meaning—have embraced ideas and agendas that have only compounded problems. I am a realist and I hope that I will always use common sense when assessing the needs of the mustang.

Mustangs are highly trainable and versatile animals. There is no question that they can become trustworthy partners for

Pages 16–17: **Wild horses on high alert—all focused on possible danger.**

people, providing years of service, entertainment, and companionship. However, if poorly treated they can become dangerous and destructive. Mustangs should never be taken for granted. To deal with any wild animal, you need a certain level of competence.

But once you have developed an understanding with a wild horse, an understanding rooted in nonviolence, you often have a very fine partner. Raised in nature, mustangs possess the kind of sharp wits rarely seen in the domestic horse. Survival of the fittest weeds out a high percentage of the herds' slow thinkers at an early age. The bright young survivors are further educated by a sociological order within the mustangs' family group, which constitutes an equine university.

The mustang must deal with the realities of life in the wild, which includes being attacked by cougars or coyotes or suffering as the recipient of stray or well-aimed gunfire, with the lead buried in muscle as a lifetime reminder. The mustang's experience of growing up in this harsh and sometimes unforgiving world ensures that he or she will either learn quickly or perish.

Mustangs miss nothing. They will notice the slightest movement within a quarter of a mile, changes in their surroundings so subtle that the domestic horse would pass them by without a glance. Their attention to detail can be a great asset or a dangerous characteristic, depending on your level of competency.

Recently I was on a mountaintop in central California with four horses and three other men. We were sitting, waiting for the ranch owner to arrive, when one of the horses lifted his head, shifted his legs, and stood with his attention directed at something off in the distance. It happened that this horse was a well-trained mustang used for working cattle on the ranch.

He stood frozen while the other horses rested a hind leg, drooped their lower lips, and relaxed their ears until they pointed sideways. The other men didn't notice anything, but I worked, trying to focus on the spot the mustang seemed to be pointing out. After a few minutes, I spotted three cows that had appeared from a draw on a hillside at least a mile away. It seemed to me that we could have spent the rest of the day there without any of the other horses noticing them.

Mustangs can work as our partners if we learn to understand them, trust them, and treat them with respect when we bring them in from the wild.

That was just what I had in mind on the cold, misty night in the Cuyama Valley, the night I brought Shy Boy in from the wild.

2

My Mustang Teachers

I was a boy in the desert learning the language of horses.

That night in the valley, as I rode in Shy Boy's wake, I felt a kinship—if not yet with him, then with his kind. We shared, in a way, a dark past. Those of his ancestors who escaped slaughterhouse trucks were conscripted earlier this century as warhorses or ranch horses, but only after being savagely broken. I was no stranger to that cruelty, for I had seen it inflicted on horses, and had had it inflicted on me.

You could say I was born on a horse. I was up there on the saddle in front of my mother for hours at a time from the day I

Two mustangs grooming each other, part of good manners in the herd and communication between horses.

could hold up my head. As a toddler I spent hours a day on a horse, watching his neck rise and fall, and the way his ears talked back and forth. I won my first riding trophy at the age of four.

My family operated a horse facility in Salinas, California, and we were dealing with hundreds of horses every day. From the age of eleven, I was mucking out twenty-two stalls a day before school. My early success in the showring led my father to believe that I would be the one to make the name Roberts famous in the world of horses.

But I had this dilemma. I watched my father breaking horses and I was repelled by the violence involved in the traditional methods he used. I watched him tie up six horses at a time in a corral, the horses evenly spaced, each one at his post. My father proceeded to tie up their hind legs so they could not move and then he would deliberately frighten them by throwing a weighted sack at their hindquarters. Of course, they would resist.

In my father's view, when the horses were totally submissive and showed no defiance, no matter what he did, then he had broken them. It took four to six weeks.

"Broken" was the right word. These horses were traumatized, and worked only out of fear. They never forgave for the pain they had suffered, never connected with a human being—out of emotional need or affection or any other laudable motive. This was tyranny.

I watched this from an early age, and I found it repulsive.

Yet this man was my father. He was an admired and respected horseman. His way was the norm, and it still is in many places in the world, more places than you might think.

What turned me against my father, as much as his cruelty

to animals, was his cruelty to me. He beat me with chains and I ended up in the hospital with my mother begging me to pretend it was an accident; she would have a word with my father, she said, change his behavior. I learned to fear him and could never fathom his desire to subordinate and humiliate those near him, whether animal or human.

At the same time, I learned I had an affinity with horses. I could watch them endlessly. Just by being passive, by allowing them their natural behavior, I began to piece together the way they communicated with each other.

In 1948 I made my first trip to Nevada. Between the ages of thirteen and seventeen, I spent time on the high ranges, south of Battle Mountain, completely alone and watching herds of mustangs so wild they had perhaps never set eyes on a human being. Tonopah, Nevada—the region where Shy Boy was born—was just a few hundred miles due south.

My desire to understand and communicate with horses was buried deep inside me and at this age was driven more by intuition than intent. Part of me was as much horse as human, and I felt drawn to make sense of the horse's physical language. I am grateful now that I heeded that instinct: I look back at that time in the desert as a critical part of my education, among the most wonderful experiences of my teenage years.

I was there, even at that young age, as part of groups sent by the Salinas Rodeo Association, to gather mustangs for use in the wild horse race, a feature of the annual rodeo. The wild

Pages 24–25: **The glossy coats of these mustangs would suggest their good health but the land available to wild horses is limited and so their numbers must be controlled.**

horse race was a kind of free-for-all in which teams of three men tried to saddle, bridle, and ride wild horses. Afterward, the horses were auctioned off.

On these annual mustang roundups—the first was in 1950—I spent long hours observing wild herds, so caught up in what I was seeing that sometimes I forgot why I was there. I was mesmerized, especially, by the matriarch. The stallion, I learned, has a role as the herd's protector, but the dominant mare actually leads the herd and makes day-to-day decisions about where to graze and water. The matriarch had a particular way of dealing with antisocial behavior. She would drive the offender from the herd, and when she released the pressure on him, the offender knew he was allowed to come back.

It's the principle of "advance and retreat." The theory goes like this: if you push a herd of wild horses in one direction for a certain distance, then lay off them and turn back, their natural inclination is to follow. Native Americans used this knowledge to capture wild herds. They would drive the herd away for a least a day and in their wake construct a keyhole-shaped structure a quarter of a mile long, using wires, poles, and brush.

When they stopped driving the herd, the horses would bend back and follow them. Several riders would then circle around behind the mustangs and drive them toward the trap in their path. Advance and retreat.

In my mind, that principle and the matriarch's disciplinary ways were linked.

If adolescent horses misbehaved—took a bite out of an elder's rump or kicked or got high-handed—they had the lead mare to deal with. She would square up to the delinquent and drive him or her out.

The offender, of course, was frightened. To be evicted from the herd is like having your death warrant signed; there are predators out there. Often elders of the herd who had outlived their usefulness would go off alone. It was a virtual death sentence, but self-imposed.

When the adolescent horse realized his predicament, he would ask to be let back into the group. He would offer recognizable signals whereby he asked forgiveness and showed submission. He would "lick and chew"—the classic mouthing action of the dependent horse asking for food from a superior. He would drop his nose close to the ground, another gesture of submission.

At this point the matriarch would respond. Where previously she had adopted the straight-ahead stare, her body pointed like an arrow at the offender, now she offered a new pose. She would turn and show her flank, change her attitude, and avoid eye contact. These were her signs that the offender had been forgiven. Permission had been granted to rejoin the herd.

This body language, I now know, is ingrained in the genetic, tribal memory of all the world's horses.

But back then I was a boy in the desert learning the language of horses. I gave this language a name, Equus, and learned how to use it. I found that I could communicate with horses, however wild they were.

By using this language, I could have a wild horse following me around as if I were the matriarch. The steps and movements—the sequence of communications in the process—were

measurable; they could be repeated. The horse's responses were predictable. Later I would discover that the method held up to the most scientific analysis of animal behavior. It was a language that could be used for communication between horses and humans.

I had learned to listen to horses.

I went on to perform this experiment in the round pen back in Salinas with the mustangs we had captured for the wild horse race.

But my first testing of these theories took place on the high desert of Nevada, without a fence or another human being in sight. My desire to be with the mustangs, to understand and learn from them, had taught me something beyond anything I knew.

My knowledge was limited. I was only a kid. Nonetheless, I did spend time with one particular mustang and caused him to stop going away from me and instead to come back. After twenty-four hours, this little mustang was following me around. These were the wildest, untamed horses of America, and here was one of them in effect inviting me to be his friend.

I stopped the experiment short to race back to the Campbell Ranch near the town of Eureka in central Nevada, which was our base for the roundup, and tell people what I had done. I was certain more experienced horsemen would embrace my ideas and try the process themselves. Surely, I thought, the approach would spread and revolutionize how we deal with horses. I can now see how this sounded like a wild tale by a

A wild mustang with cockleburs in his forelock.

young man with an overactive imagination. It was laughed off.

When I returned home to Salinas with the captured mustangs, my father and his friends also found my tale amusing. They slapped their legs at its mention. Given the episodes of violence I regularly suffered at the hands of my father, I learned to keep quiet.

The following year, 1951, when I was again allowed to help gather mustangs for the rodeo, I returned to the high desert with the aim of doing all I could to consolidate my experience of the previous year.

This time I chose a male, about three or four years old, agile, fit, and intelligent. Within twenty-four hours I had reached a point where I could ask him to come back to me. By aligning my body perfectly with my mount's body—eyes straight ahead, the axis of my shoulder and hips matching my horse's—I could stop the young mustang and have him follow me. This time I could get close enough to stroke him. Leaning over and touching this wild animal was an exhilarating experience.

The wild horse had decided that I was no threat. It was as though I were one of his family members scratching the top of his neck. I slipped a loose rope around his neck and schooled him to "lead," so he would follow alongside my saddle horse. He was not particularly resentful of that; as far as he was concerned, we were both on the same team.

I took a surcingle, a beltlike device that lets horses get used to the notion of a saddle, and placed it over his back. Not wanting to risk going under his belly to catch the other end, I had fashioned a piece of wire into a long hook, which I now used to bring the surcingle under his stomach, so I could finally buckle it around him.

During much of this time I was holding my breath. I was in wild country and risking my neck, but it seemed the most natural thing in the world: the mustang, the open desert, and me.

He didn't buck. He was goosey, but he didn't buck. I asked him to accept a snaffle bit, and he did so. Later on, I worked from the ground, putting a western saddle on him. When he felt the weight of it and the stirrups slapping him, he did try to buck it off now and then, for a few seconds at a time.

But he wasn't traumatized or overly frightened. I had asked him to accept quite a number of new experiences that day, but I recognized I wasn't gong to be able to ride him until he had relaxed by a few degrees—I was a long way from help if I were bucked off and my head hit one of the stones that littered the ground.

It had been a successful experiment, and I called it a day. I removed the tack and turned him loose. I could hardly believe that I had developed an intensely personal relationship with this wild horse. But there was not a soul to witness it and I was dying to tell someone.

When I returned to the Campbell Ranch, I told the hired hands there what I had done. Like my father, they laughed. They saw me as a young man who had made up this whole episode out of excessive enthusiasm and a vain desire to promote his abilities.

I was determined to come back the following year, in 1952. This time I would give myself a day or two more, to see if I could actually ride a mustang back to the ranch. There would be no doubting it then.

At home, in secret, I began using this newfound technique, which I would call "join-up," to "start" the mustangs—to intro-

duce them gently to bit, saddle, bridle, and rider. After the rodeo had finished, there were about a hundred head to be sold off, and my brother and I decided to assign them some extra value by giving them basic schooling. He broke his share, using conventional techniques, but I started mine using join-up. Given the number of horses, I was already beginning to refine my technique. And it struck me that the new way was a lot quicker than the old way.

The following year I was better prepared. By then I knew much more about what I was doing and had allowed myself the time necessary to achieve my aim—join-up in the wild, then saddle, bridle, and ride the wild horse.

I found a strong bay colt probably four to five years old, showing a lot of Andalusian characteristics: high action in front, feathers on his fetlocks, and a muscular neck and shoulder. He had large, black eyes with a light burning in each of them. An experienced horseman or horsewoman would say he had an intelligent eye. His entire appearance appealed to me.

As I cut him away from the herd, he was a magnificent sight. He arched his neck, held his nose up, and stuck his tail right up in the air so the hair flowed down over his hips. His tail was so long and full that it created a black veil over his haunches down to his hocks. He cantered away and I followed.

I was struck by his short, powerful stride, so unlike that of the Thoroughbred. As I followed, bending him in different directions but essentially driving him away, we settled into a pattern. I knew what to look for, the licking and chewing, the head dipping to ground level. My experience in the round pen at home had taught me a lot more about interpreting the language of Equus.

For one thing, I was quicker. Before twenty-four hours had

passed, I could drop a rope around the horse's neck and lead him around. He was wild, but seemed to be trying hard for me. When I got to the stage of wanting to drop a longline around his rear quarter, he kicked with a purpose. Longlines are thirty-foot-long flat ropes, like long reins, which run from the horse's head, through the stirrups, and back to the trainer on the ground. The aim is to teach the horse something about steering—turning left, right, and backing up.

Each time he kicked I found myself whispering, "Hey, buster," so then he had his name. From that moment I would refer to him as Buster.

After schooling to get him used to the longlines, I could literally see him decide not to kick. I would then rub him on the hip bone to congratulate him. The surcingle was on him at around the twenty-four-hour mark and he didn't buck. After about thirty hours I was leading him around, saddled and bridled. He kicked at the stirrups a few times, but his bucking was limited to a few crow hops, nothing more.

On the afternoon of the third day, I put one foot in the stirrup and lifted my weight onto it. He circled around, and it seemed a good moment to swing my leg over and sit on him right then.

First, though, I took some precautions. I scouted the area on my saddle horse to find a patch with fewer stones so that if I was to be thrown, I would be less likely to be injured. I also had twenty-five feet of rope coiled in my belt and linked to Buster in a kind of halter arrangement, so if I was bucked off, I would have a chance of holding on to Buster.

On the morning of the fourth day, I rode him. There was no round pen, not a fence in sight to contain him. Yet here I was,

putting a foot in the stirrup and lifting myself on. We were pin-points in that huge desert where the sky arched over us, yet I felt we were at the center of the universe.

Throughout most of that afternoon, I rode Buster. I took care only to ride him for a few minutes at a time and made sure I mounted and dismounted on different sides, so there would be no surprises, no new cause for him to buck. I also had him get used to leading my saddle horses, for it was my intention to ride him the twenty miles back to the Campbell Ranch, where a large con-tingent of cowboys would see that I was not making this up.

For most of the way I rode a saddle horse to keep Buster fresh. Two miles out, and with no little sense of anticipation, I switched to Buster. I rode him at a trot, leading the other horses. I felt a sweet sense of accomplishment as Buster marched into the ranch like he'd been doing it all his life.

A group of men were doctoring calves in a corral, another few were working on a generator motor over by a barn, and others were coming out of the bunk house. They looked me over and asked, "What's going on?"

When I told them what had happened, I expected to see their expressions change, to hear questions. Instead there was only skepticism.

"You must have taken an already broken horse and salted the herd," said one cowboy.

"Look at his feet," I countered. "He's never had a shoe on. You can see he's an out-and-out mustang."

Winters can even be hard on mustangs, among the hardiest horses on the planet. But these two horses seem well-equipped with their thick coats.

"I can see he must have been out there," he said, "but you were lucky enough to find one that someone's gotten to first."

In that instant my morale plummeted. I felt angry and frustrated. This had been a nerve-racking four days for me. I had achieved what I thought was a significant breakthrough in the way we deal with horses. When we returned from the high desert to Salinas with our mustangs, two or three of the hands were telling people they had seen me ride in on this mustang. And I myself spoke of it.

But in the end no one gave it credence. Though I continued to use join-up, I put the whole high desert experience out of my mind.

Some forty years later, the notion of gentling a mustang in the wild resurfaced in a way I could never have predicted. In the fall of 1996 I happened to be in London on tour, demonstrating join-up to audiences all over Britain. Executives of BBC television, who had made the original documentary on the process of join-up, had contacted me. The film had enjoyed considerable success and we were discussing the possibility of other programs.

We met in London in a massive marble building they call White City. The building is like a self-contained community where decisions are made on programming.

"What's next?" someone asked. Was there another program to be made with Monty Roberts? Did I have any new ideas?

I hesitated. This wasn't a small idea at all. Nonetheless, I pressed on.

"There is one experience I've always wanted to relive." They waited patiently.

"When I was seventeen I did actually achieve join-up with a mustang in the wild."

There was silence.

"What I'd like," I went on, "is to re-create that event, to take the principles of join-up, do it in the wild with no round pen and with a totally wild mustang. I've done it once before and I think I can do it again."

They asked, "What makes you think it'll happen the way you want it to?"

I explained that the principles of join-up were given to me by wild horses living in natural, unaffected conditions on the high desert in Nevada. What works in round pens before audiences all over the world would work in the very location where I learned it.

The BBC seemed interested, but quite naturally could not commit to the idea. It was a world away from their own. There was a lot of risk from their point of view: the enormous expense of travel, film crew, director, and equipment, to list a few of the obstacles. The weather, snakebite, injury or failure on my part—any of those things could see them spending a boatload of money with nothing to show for it. For them it was new territory and the financial risk would be all theirs.

I left feeling fairly sure they would reject the idea. But I also recall looking back at that building and thinking what a difference it could make in the treatment of horses if by some miraculous turn of events the BBC would agree to document my story and my method. I stood there with all sorts of thoughts running through my brain. Could I really do this? At sixty-two,

I had lost a step but gained in other ways. I weighed the differences. As a teenager I had been a fit athlete but had been inexperienced in gentling horses. Now I was an older man, with a lot of broken bones along the way, but I could count on a lifetime of working with horses.

The idea for the documentary had been launched, but it was now out of my hands and I would dwell on it no more.

In January 1997, when I got back to Flag Is Up Farms, my home in California, there was a message waiting for me from the BBC. If I was willing to try it, they were willing to take the risk.

Shy Boy, here I come.

3

IN SHY BOY'S SHADOW

I had been up for thirty-six hours, all of it in the saddle, and done some of the most arduous riding of my career. I was exhausted.

The first hurdle was just getting to Shy Boy. The mustang, a part of our heritage, is an endangered species in America.

Consequently, and quite properly, the mustang is protected by federal law. An Act of Congress forbids anyone interfering with them in any way. They are under the authority of the Bureau of Land Management and it is illegal to go near them in their wild state without authorization to do so.

My only hope for acquiring a mustang was to adopt one. The BLM runs a system of herd management, whereby a certain number of mustangs are taken off the open range, to be offered up for adoption to approved homes. During these roundups, helicopters are often used. Their use has sparked controversy, but for me the key issue is not whether they're used but how they're used. A good pilot can do the job well; a bad pilot may run horses into the ground.

The horses are not sold, since the profit motive would com-promise the situation (although after a year—and following BLM inspection—owners may apply for full ownership). Names are put into a hat and if you are lucky your name comes up and you can adopt up to four horses at a time. However, this mustang lottery only happens occasionally, so when I got the news that the BBC wanted to make the documentary within three months, my heart sank. Was there an adoption event soon enough? We would need a lot of luck. If we could adopt a few of these mustangs, we could then place one of them back in a wild environment where we could conduct our on-screen experiment.

On the very day I was weighing all this at our farm, one of my students rushed up to me. Flag Is Up Farms typically has five students on the premises to learn horse gentling.

"Monty, look, there's an adoption event."

"When?"

"Tomorrow!"

There was the luck. It seemed like this was fated to happen. Next day we jumped in the truck and headed to Santa Margarita, a small town about seventy-five miles north of the farm.

It is important to realize that a mustang captured off the range is no less wild in the corral; quite the opposite. His adrenaline would be sky high. Stallions like Shy Boy are gelded; the mustangs are kept in holding pens, transported to the adoption site: The whole experience would have every horse in full flight mode and less easy to settle.

I was after males, not females, because they were likely to be pregnant. I wanted three horses: a first choice, as good-look-ing as possible for the cameras, plus two backup horses in case

some misfortune should befall my first pick. And I wanted all of them three to four years of age.

Among the 220 horses up for adoption that day, only 20 fit my criteria. We put our names in a hat along with fifty-five other people. Once more, the whole enterprise hinged on fate and more good luck. We waited, my students and I, while the names were called.

By number forty-nine, my name still had not been called. Fifty-two, fifty-three, fifty-four . . . The last name out of the hat was mine.

Our choices seemed limited. But as I looked into the corral that day, one horse in particular caught my eye. He was more handsome than the rest, with a brighter spark in his eye. When he moved to avoid human contact, he did so more quickly and fiercely than the others. I could see manifested in his build and in the way he carried himself the classic Spanish ancestry, the proud, strong heritage of his forefathers.

He was looking at me out of the corner of his eye and he wanted nothing to do with me. It was as though he was saying, "Stay away. I'm wild and I'm going to stay that way. I don't know what's going on here, this is outside my experience, and I sure don't want to find out." Here was a character to contend with, but equally, one anyone would be proud to know.

This was Shy Boy. I didn't have his name yet, but we were about to start a relationship as close as any I have had with a wild horse.

I made note of the small plastic number tag he wore at the end of a nylon rope around his neck. Number 212. Then I moved on to identify the backup horses and to write down two more numbers. Given that I had been the last name out of the

hat, I would be lucky if any horses were still available when it came to my turn.

Again, our luck held, and I got the mustangs I wanted. Someone was watching over me.

Back home at Flag Is Up Farms, the mustangs entered the environment of the domestic horse. They were worried, confused. When they heard the sounds of other horses calling out to one another, they ran to a corner of their enclosure and huddled together, lowered their heads, and walked as stealthily as possible. In their view, the domestic horses were breaking the rules; they were attracting predators from miles around with their unnecessary calling. I had witnessed this phenomenon many times before, and now my students were seeing firsthand the difference in the culture of the wild horse.

My first concern was to get Shy Boy away from the farm. I did not want him around people and I certainly did not want him around me. I had discussed with the BBC the measures we ought to take so that viewers of the film would know we were dealing with a wild horse. We came up with the plan of attaching an independent referee, one who would monitor the mustangs until the start of the project. That person would ensure that the horses remained untouched and away from humans.

Carol Childerley, an animal rights activist with a degree in wildlife care and management and a volunteer with the Santa Barbara Wildwatch Association, took charge of all three mustangs. She made arrangements to transport them up to the high desert of the Cuyama Valley, almost two hours north of our farm, there to be mixed with a privately owned free-ranging herd on Claudia Russell's ranch of thousands of acres. Carol would check them regularly.

Once mustangs have been adopted, they cannot go back onto BLM land. The adoption charter requires that they be kept on private lands. The ranch we had the good fortune to be using was actually very close to the mustangs' natural habitat. The only reminder it wasn't virgin territory was a fence line every six to twelve miles, or maybe a windmill on the horizon.

Now that we had recruited Shy Boy, I felt our major organizational difficulties were over. And the BBC had assigned to the project a director known for his documentary work.

When the BBC called, it was January 1997. The rattlesnakes come out of hibernation in April, and the risk of fire increases rapidly, so the filming would have to take place before then. I also had to plot for some moonlight so I could have night vision. The ideal time would be the last weekend in March, which happened to be the Easter weekend. All we had to do now was to prepare our saddle horses and equipment and wait.

One lingering doubt centered on me. What was I letting myself in for? My back was in a poor state: Some fifteen years beforehand it had been welded together along five vertebrae and the soft inner cushions of five discs had been removed. Could I really ride for two solid days, night and day, through the wild extremes of temperature you can get in the high desert? And could I do all this and still accomplish the goals set for the project: achieving join-up in the wild?

I set those doubts aside. I had been given the opportunity to film the process and the funds to mount such an expedition. What slammed home to me was how much it meant, how widely a film could disseminate the message of nonviolence. There was no way I would turn this down. It was all systems go.

The Cuyama Valley lies between the Sierra Madres and the Caliente range in California. There is a rugged beauty to the place, especially at sunset when the shadows on the ridges of the mountains grow long. Look one way, and it seems the mountains are looming over you. Turn and you see rolling hills. Look another way and the horizon seems to stretch on forever before finally touching the sky.

Anyone coming here for the first time would be struck by the colors of the rocks: ocher, burnt orange, slate gray, chalk, and cream. A complement to these colors is the high desert flora. Yucca is a particular favorite of mine. It flowers annually with a white, almost luminous bloom held high on a stem and pollinated by moths in the moonlight. It's known as the "desert candle." The ground in which it takes root is a combination of granite and sandstone, quartz, feldspar, and fine loam.

Geologically, the area is unusual. At its eastern end is the San Andreas fault, which here changes its direction about 25 degrees to a more northwesterly orientation. The shifting of vast underground plates causes the mountains to tip slightly toward the sea. It sounds complicated, even perilous, but this gyration of the earth's crust is fairly civilized—except for the occasional earthquake.

The Cuyama Valley sits in what is known as a transverse range—one running east to west between the dominant ranges running north to south. The valley is dry because it sees many

A strategy session Saturday night in the Maverick Saloon in Santa Ynez, California, involving Pat Roberts, Monty, and one of the BBC film crew.

hours of sun. In March 1997 the valley was parched. Any movement on the dirt roads—from trucks, horses, or cattle—sent up man-high clouds of dust.

Shy Boy was taken into this valley, along with his two backup mustangs. All joined the privately owned herd. The mustangs would feel at home out here: it was not actual wilderness, but felt like it. Some of the ranches in the area are private holdings as big as 40,000 acres.

I was ready to show that join-up would work even out here—without a round pen in sight. Just myself, the mustang, and room to run.

For this task I would have with me three of my horses. Dually, of course, would be there. He has been schooled so precisely that his movements can be guided to within an inch, using only the lightest of touches. I would use Dually for the last stage, for join-up, when I would need a horse that could accurately imitate the stance and movements of a wild-herd matriarch.

I also had The Cadet, a dark, handsome ranch horse with a big trot: he trots at around twelve and a half miles an hour. He would be useful when things settled down after I cut Shy Boy from his adoptive herd. I would then have to ride for many, many hours.

Third, I had Big Red Fox, a tall retired racehorse. He seemed to me fit for the second phase of the job.

The plan was to use The Cadet to cut Shy Boy from the herd, then at dusk switch to Big Red Fox for the moonlit

Freddy at the Maverick Saloon.

period—allowing Shy Boy time to rest and eat. Next morning I would switch to Dually for the delicate, highly charged moments of join-up. All three horses had very fit cardiovascular systems. I myself had been trotting a horse three to five miles a day, trying to get in shape. The ropes, saddles, tents, and the entire equipment list for the expedition had been collected, gone over, and rechecked dozens of times.

Five ranch hands were available to me as wrangler crew. One of them was Caleb Twissleman, a sixteen-year-old boy who was to shadow me personally. I like everything about Caleb: he's a phenomenal athlete, rider, and roper who has won about twenty saddles in western competition. The other four members of the crew, including Cathie Twissleman, Caleb's mother, were to take Shy Boy's adoptive herd off in the other direction once we had cut him out.

We had taken care in selecting the site for the base camp—high atop a hill in the center of the 1,200 acres allocated to us, near good grazing areas. This would be helpful when it came time for Shy Boy to eat. Base camp had to look after quite a population: there would be the BBC crew to feed as well as the wranglers and various other assistants.

All of us would have to step carefully. The rattlesnakes, we were warned, had come out early. They strike at body heat, and when they first emerge from their dark winter dens, they're cranky and unpredictable.

The film shoot actually began Saturday, the twenty-ninth

The free-ranging herd in the quiet splendor of the Cuyama Valley at dawn Sunday morning.

of March. First shots were not of a beautiful dawn and a wild mustang in flight through sagebrush. They were, instead, of the Maverick Saloon in Santa Ynez, where a lot of fancy dressers were line-dancing to Art Green's band. The film crew was happy to find there an old-timer named Dutch Wilson.

"Maunteee's absolutely crazeee," he drawled to anyone who would listen. "He stands as good a chance of being killed outright." Then he added, "So who is it wants to go with him?"

Others joined in. "You want my opinion? It's not going to be easy. Lot of people bin killed, tryin' to break mustangs."

"Tell you something about mustangs. The most dangerous part of him is his front feet. His back feet'll hurt you, but his front feet'll kill ya!"

"You see a lot of cowboys round here with broken teeth."

"And the bones to go with 'em."

Still, I felt relaxed and as ready as I would ever be. If I failed to accomplish my dream, I thought to myself, it would not be because I hadn't had a chance at it.

I left the saloon around nine P.M. and drove for approximately an hour and a half to the high desert. There I slept in my trailer for a few hours. It was bitterly cold, a warning of what lay ahead.

Then it was daylight, Easter Sunday.

The first thing I did was prepare my body for the epic ride ahead. To the insides of my knees, thighs, and calves I stuck what looks like a second skin. Made by 3M and called Tegaderm, it's a tough, transparent membrane used by marathon runners for protection against blisters. But because the membrane sticks like glue, I first ran a razor over the areas to be cov-

ered to reduce the pain of removal. The hairs on my legs would later thank me.

Over this tape I pulled on long johns, trousers, and chaps. I was prepared to ride for a day, a night, and then probably most of the next day as well. For my bad back, I took anti-inflammatory tablets, and of course I was strapped into a back brace. To protect myself against cold, I donned five layers of clothing, although this would still leave my hands and face exposed.

I have never worn gloves in my life; they never seem to make my hands any warmer and I have never gotten used to reining and roping with them on.

At first light the ranch hands gathered to clear cattle from the area, leaving it open for the horses. Last-minute adjustments were made as to who was riding what horse, and then we set out, moving west alongside the Cuyama River. My group consisted of Pat Russell, Cathie and Caleb Twissleman, Barney Skelton, Scott Silvera, and me. Somewhere out there, Shy Boy was waiting for us.

This is big country, in every sense of the word. It's where they come to make Marlboro cigarette ads. The early sun brought the place alive. The bees set out from their white-boxed hives for the sages, and the birds grew animated. The light struck the sycamores that line the riverbeds in the valleys, along with the yucca and cactus that grow here in abundance.

Pages 52–53: **Monty on Dually and Cathie Twissleman on Splash start to cut Shy Boy from the herd. He's the third horse from the left, distinguished by the white stocking on his left hind leg.**

We met up with the free horses more or less where we expected to, about a mile on. Shy Boy was there, his head high, one of the crowd to observe our approach, yet he stood out because of his looks and his inquiring nature. He seemed more or less assimilated into his new family. He was lean, fit, in prime condition.

Of course, he had no idea of what was about to happen or the resources already expended. He could not have known about the film director and crew, the wrangling operation, even a helicopter—all centered on him.

The BBC had also hired a veterinarian and animal behaviorist, Dr. Robert Miller, to observe and comment for the camera on what I was doing. He was also there to verify that Shy Boy was as wild as a deer and to give his seal of approval to the shoot: but only if no harm came to any animal, from pack horse to Shy Boy.

Carol Childerly, who had now been in charge of the three mustangs for nearly two months, was also on the scene to verify that Shy Boy was the same wild mustang adopted in Paso Robles. She carefully observed his movements and reported to the BBC: he was a true wild horse.

Shy Boy, just to the right of Cathie and Monty, in flight.

Pages 56–57: **Shy Boy is now out of the herd and on full alert.**

Page 58: **Monty appears to be enjoying the challenge—at least in the early going.**

Page 59: **Shy Boy makes an attempt to rejoin the herd, but Dually is on the job.**

The photographer Christopher Dydyk took up a position just behind the film crew with tripods, a Hasselblad camera, and lenses the size of stovepipes. He was prepared to document the event with still photographs, and he did just that—the proof lies in the marvelous works of art that grace these pages.

The film crew moved into place, and I gave my instructions. The six of us would ride into the herd and Caleb on Tari and I riding The Cadet, would cut out Shy Boy to take him east toward the open country. Like his forbears, he would fly when he felt himself apart from the herd. I also knew I would have a job keeping up with him. Caleb would follow me, leading a pack horse. Meanwhile, the others would keep the rest of the herd going in the opposite direction, west. It was best if they were out of the way as soon as possible. I didn't want Shy Boy to feel any pull toward them. I wanted them out of sight and mind.

We rode in and split them like snooker balls. The herd took off in one direction and I drove Shy Boy in full flight the other way. The film shot from the helicopter shows a sleek, fast mustang, fluid and youthful, followed by a well-padded old cowboy galloping behind him. Shy Boy's vigor, strength, innocence, and that spirited element of his character were now all the more evident set alongside my old age and the infirmities that come with it.

Pages 60–61: **Cathie helps Monty as Shy Boy tests them.**

Separate now, Shy Boy and Monty begin the journey.

Pages 64–65: **Shy Boy heads east for higher country.**

What I had going for me was my experience with horses. In terms of my technique, I was now sending the horse around the ring, causing him to go into flight. In terms of the psychology of the wild herd, I was the dominant mare running at the adolescent, driving him away to show I was displeased and asking him to pay me some respect.

What I had not reckoned on was the level of panic induced by the helicopter. Aerial shots were essential to capture the experience on film, but because mustangs are regularly driven and captured in the wild by aircraft, the horses are frightened of them. Shy Boy was no exception. He was running hard, with me in pursuit. I have some regrets that this part of the filming did not go as planned.

I wanted to wave the helicopter back some distance away. But it was no easy task, in a full gallop, trying to signal the helicopter while using a radio tied to my saddle bags. The pilot couldn't see my signal, the white T-shirt I had earlier stuffed in my jacket and was now waving in the air. Much of the time I had to be up out of the saddle, standing in the stirrups, to save The Cadet's back. Never mind my welded spine. I was unsure of the ground and a fall at that speed could have been serious.

The helicopter eventually got the message and stayed farther back. We eased off to a workable pace. Then came another

Pages 66–67: **Shy Boy on the fly headed east. Monty and Cathie stay with him.**

Headed northeast. Against the magnificent backdrop of the Caliente Mountains, Shy Boy gallops off with Monty and Dually close behind.

challenge. We had covered nearly twelve miles and were head-ing toward the first fence, a barbed-wire cattle fence. When they're in flight like this, mustangs do not see fences well. It was critically important to keep him safe.

The task was to circle his flank and head him off. I picked up the pace, and, moving at a slight angle, managed to bend him around, well short of the fence.

That afternoon, Shy Boy led me to every water hole, stream, and cattle trough in the area. It reminded me how good mustangs are at finding water in the high desert, with its long dry spells.

There were areas of an acre or so where ground squirrels had burrowed, and I would watch Shy Boy skip around these holes, barely breaking step. I had to be very careful indeed not to put The Cadet straight into one of them and cause a fall.

As I rode I pulled beef jerky from my bag and sipped water from the canteen. I talked to myself and to Shy Boy, encourag-ing us both to keep going. It was one wild ride.

At dusk things settled. We slowed dramatically. It was not my intention to pursue him through the night, only to follow, keeping tabs on him. Shy Boy needed to rest, eat, and drink. So did I.

By this point I had been riding for seven or eight hours at a fairly good clip. The Cadet also had to rest, especially since the helicopter had pushed us more than I would have liked. We had

Northeast and bending. Like all mustangs, Shy Boy has extraordinary endurance and he seems to run effortlessly.

counted on the strength and fitness of both Shy Boy and The Cadet, and we had come through Phase 1.

Now for the tricky part. Night.

Caleb rode up alongside and handed over my night horse, Big Red Fox. He took The Cadet back to the camp, where my wife, Pat, groomed, fed, and watered him. She bandaged his legs with poultices to keep any heat from building up and gave him electrolytes to replenish his fluids.

I watched Shy Boy take on water and graze. He would keep an eye on me and occasionally dart off, but then I would trail him. He was disconcerted, asking me for an explanation. But he wasn't asking to join-up with me. Not yet.

As it got darker Shy Boy slowed and I was pleased to let him. I began to watch him more closely, assessing his character. What sort of animal was he? From his response to me during these last minutes of dusk I judged him to have an impish quality. That spring to his stride added to his appeal.

It was during this night that the mustang became Shy Boy. Big Red Fox seemed able to follow the wild horse, even after a cold heavy mist stole the moonlight. If we were indeed following a ghost, he had mysteriously left a trail for my horse to follow.

Like a mantra, a little song formed in my head. I uttered its phrases aloud to keep my mind focused and off the cold and to let the mustang know by my voice that I meant him no harm. The song went, in part, like this:

Headed east. Shy Boy is moving at a good clip.

 S H Y B O Y

Hey, Little Shy Boy, where are you going?
Stop trying to hide from me.
I'll be here come morning when there's light for me to see.

Hey, Little Shy Boy, don't be afraid.
I'm not going to hurt you. That's a promise made.

Fifty years back, and to your kin.
My word was, it'll be better than it's been.

I told them I'd leave the world a better place
For both you horses and the human race.

Hey, Little Shy Boy, settle down.
Eat, drink, stop running around.

You think I might hurt you, but that can't be.
Because of the promise I've made, you see.

No pain to horses. Now that's my goal.
It's my life's work. I'll take this role.

Shy Boy, Shy Boy, don't be so shy.
I know we'll be friends in the by and by.

By midnight or so, Shy Boy had filled himself with good grass and had even eaten some alfalfa the men had thrown out along the trail; he seemed to assume a more relaxed and restful attitude. I was happy to oblige him and relieve Big Red Fox of my weight. The three of us would stand and rest for fifteen-minute stretches.

My spirits were still high: I was, after all, doing what I had longed to do. My body, though, was aching and tired after being in the saddle for eighteen hours. The cold had reached my bones, and my nose and ears felt as hard as icicles. Still, as long as the sturdy Big Red Fox could track this mustang, I was hopeful.

Around four-thirty A.M. a glow of light appeared in the eastern sky. With that first hint of sunup, I noticed a quickening in Shy Boy's tempo. With dawn, he picked up the pace still more. Shy Boy was up to a trot. Luckily the fog was gone and I could see him more easily now. Big Red Fox kept him in his sights.

Shy Boy's head was high and thrust forward as he drank in the smells of a new day. He seemed to be headed for that glimmer of light at the rim of the world. He was off to my right about a hundred yards and heading east when he broke into a canter. Once more I had to grasp the saddle horn and raise my weight off Foxy's back. It was only then that I noticed the many blisters on the palms of my hands and on up the fingers from the day before. All that time gripping the saddle horn had taken a toll.

As the canter continued, I started to grow frightened. I was beginning to see the bottom of my endurance, and if not for the brilliance and generosity of Big Red Fox, I am not sure I would have made it.

After five A.M., still in dim light, Shy Boy settled back to a reasonable pace and I was able to encourage him to bend in a westerly direction, toward the camp. Trotting comfortably now, I used my radio to alert Pat and Caleb that a change of horses would soon be in order.

Just to the south of camp, I spotted Pat and Dually on the ridge off to my right. We were now within fifty yards of camp. Was this Shy Boy's sense of humor? Was he aiming to take me

back to my species, since it looked to him as though I had had all I could handle of the wilderness?

Shy Boy, I would later learn, wanted water, and there was a trough north of the camp. The fact that he had come so close to camp en route was proof of a dramatic change in demeanor: he was still wary of people and trucks, but he seemed possessed of a new calm. It was as if he had waltzed into the kitchen.

My daughter, Laurel, told me later they were all amazed when they saw Shy Boy, then the top of my hat, appear over the hill. She and Christopher Dydyk had been discussing the difficulties I must have faced with the fog. The weather and tension had clearly affected them all.

Quite a cheer went up as they sighted us. Laurel then shouted out, "You look like hell, Dad."

I growled back, "I feel like hell."

I didn't know it, but my face was black, caked with sweat and dust. The wind had flattened my hat brim right back to its crown: I looked like the last man in a pretty sorry posse. Laurel ran alongside and gave me a Diet Pepsi and a piece of beef jerky. I took them gladly.

I caught sight of Pat, and our eyes locked for a few moments. I saw her eyes fill with emotion and I felt the same.

"My God, Monty," she called out. "Are you all right? Can you make it?" It hit home how near failure I really was. But as I left camp, and as the animated voices faded behind me, my morale lifted. This contact with my family and crew after the uncertainty and isolation of the night had been a boost.

It was now around five-thirty in the dimmest of light, and I could see the headlights of the BBC vehicles kicking up dust on the dirt road a few miles east of camp.

The director's voice crackled to life on the walkie-talkie and I told him we were running according to schedule. I had been in the saddle for twenty-four hours but was still on course, with Shy Boy in sight. I gave the director a fix on my position so he could do any filming he wanted. We were in a slow trot now, and Shy Boy was taking me north to the water trough. I stood off about fifty yards as Shy Boy lowered his muzzle into the large galvanized steel container, around a yard deep and two by four yards in area.

I waited. Then, I waited some more. Unless he was going to drink the whole tank, something was not right.

I edged closer and Shy Boy headed off, but no water dripped from his muzzle. Then I saw why. A thick layer of ice covered the surface of the trough. It had indeed been a cold night.

I dismounted and took a fencing tool from my saddle bag and broke a large hole in the ice so that both horses could drink. Big Red Fox took his fill and then we stood well back to allow Shy Boy his turn. He walked to the trough keeping his usual wary eye on us, and as he drank I had time to reflect on the night I had just come through.

The beauty of it is something I will never forget: the deep dark silhouettes of the mountains at dusk. The moon hanging there, like a "night light." Comet Hale-Bopp putting on a meteoric light show for Shy Boy, Foxy, and me. The misty veil sent up by the Pacific Ocean to add to the wonderment of it all. And Big Red Fox guiding me ever onward.

And then there were the other things I cannot forget. The bitter cold and driving winds. The holes in the ground that chilled my blood but seemed no obstacle for a pair of wonderful equine athletes. The blistered hands. The back now screaming for rest.

On balance the positives of that long night far outweighed the negatives. It was after six A.M. by then and the sun was pushing the darkness across the horizon. The BBC crew were setting up their cameras about a quarter of a mile to the west of me. I felt a surge; it started in my toes and went quickly to the top of my brain. This is what they call a second wind. I had the sense that the worst was over.

I watched Shy Boy flicking his nose on the surface of the water, as if to say to the liquid how much he loved it and always wanted some within reach. For some reason, what entered my mind just then was a concept I had learned about in psychology classes a long time ago, "catastrophic bonding." It is what happens to plane-crash survivors who spend hours together on the top of a mountain.

Shy Boy, I am sure, had catastrophe in mind as he was being tracked by a human and was considering the potential for dying in the next minute. I was going through something similar: rejoicing in my mind but terrified that my body would, at any moment, betray me. When two people in a car, even strangers, are involved in a near-fatal crash, the impulse is to embrace the one who shared that experience, one that called on the ultimate in inner strength. It often results in deep and lasting friendship.

Shy Boy evidently was convinced I was not worth that kind of commitment, but I was well on my way to reaching that feeling for him.

Monty, now on The Cadet, at sundown on Sunday as the cold descends on the valley.

After he had grazed a bit, we started work again. The cameras were once more recording our every move. By eight o'clock I found I could square up to him, look him in the eye, and cause him to stop. Then I would pull up, break off eye contact, and rein Big Red Fox away from him. We were entering into the join-up phase.

The mustang had learned that flight wasn't going to work. He was going to have to deal with me, ask for my help, and enter into a dialogue. It was up to me to listen, to read the signals, and to show that I understood his language by the speed and accuracy of my response.

At this point I needed Dually. I felt I was no more than an hour or two away from join-up and I wanted him to make the accurate movements necessary to complete join-up effectively. I squeezed the button of the walkie-talkie and asked Caleb to bring me Dually.

Big Red Fox was taken back for a well-earned rest. He had been a godsend, that horse. I was as proud of him as the parent of an honor student would be. He had gone to class, quickly learned his lessons, and passed all tests with flying colors.

Dually, of course, was fresh, even playful, in the beginning. He wanted to buck and fool around. Shy Boy, I could see, was

On Monday morning Shy Boy finally stops running away and settles down for the beginning of the join-up process.

Page 82: **Join-up achieved, Shy Boy now follows Monty of his own accord.**

Page 83: **Monty on George.**

put out by the exuberance of this new horse. It was fifteen or twenty minutes before things settled down.

Now I could speak the language of Equus.

This mustang would put an ear on me, constantly rotating it in my direction, and lower his nose. In the same instant I would have Dually break away, retreat from him and ease off the pressure. That was Shy Boy's reward for going soft and for accepting me. But the instant the mustang started going away, we were on him—advancing, becoming the aggressor. Advance and retreat. Shy Boy pretty soon figured out what he had to do to stop this man and horse from coming at him. We were speaking his language and it soon led to relaxation, return, and cooperation.

The cameras were rolling. I was ecstatic. Experience has taught me that after achieving this level of communication the outcome is certain. Within forty-five minutes, I could walk in circles around Shy Boy and have him bend to follow me. He was showing that he trusted me.

At ten A.M. I asked Caleb over the walkie-talkie if he would come up and approach the other side of Shy Boy. Caleb came promptly.

Slowly, deliberately, I leaned over the saddle horn and rubbed the mustang's neck. I told Shy Boy how nice he was to decide to be with me. No ropes, no pain, no submission.

Pages 84–85: An angry Shy Boy (note the ears flattened back) seems to be saying, "I don't want you to put the rope on. Please go away."

Caleb Twissleman on Tari and Monty on George look on as a tired Shy Boy offers a gesture of submission (note the mouth).

Dr. Robert Miller, the animal behaviorist hired by the BBC, observed as he looked on, "This is an animal that could kick a fly off the wall. Yet, of his own free will, he is accepting a human touch." For me, this was a great privilege.

I felt a deep sense of triumph, tinged with exhaustion. Certain phrases began to explode in my brain as if announced by a messenger. "I am alive. I'm relatively okay. I'm riding Dually. I've completed twenty-six hours in the saddle. I've got a mustang standing with me. I've achieved join-up."

That moment ranks with other landmark incidents in my life: marriage, the birth of three children, the day I turned away from my father's traditional ways, and the first time I saw my champion horse, Johnny Tivio.

I felt a surge of renewed energy.

The next step was to drop a loose rope around his neck and school him to lead alongside Dually. Shy Boy didn't like this new idea. But when he went into flight, I didn't try to stop him. On the contrary, I agreed with him. "Go on, run away," I told him, "but don't run a little, run a lot." With the rope still on him, I drove him away until he was asking to come back. It didn't take long until he trusted me and was leading comfortably.

Most of that morning was taken up with join-up, putting the rope on him and leading him with Dually. Then came a bit of rest and a bite to eat. The animals were allowed to graze. Delegated to watch Shy Boy, Cathie Twissleman reported that

Shy Boy is now clearly more at ease in Caleb and Tari's company.

he showed no sign of flight. He was content to remain near the domestic horses.

In the afternoon I swapped Dually for George, a solid ranch horse. We spent most of the afternoon schooling Shy Boy in big circles, leading him from George. Once in a while a spark would ignite and Shy Boy would spook, but his periods of trust and cooperation were lengthening steadily.

George did his part. He's a thick horse with a low center of gravity, which made it easy for me to rub Shy Boy over the back, ribs, and hips. Shy Boy quickly became comfortable with George and this allowed me to massage Shy Boy from head to hip, from wither to brisket.

At around four in the afternoon I decided that we had progressed far enough for me to put a surcingle on Shy Boy. I remained on George, slid the surcingle down the offside of Shy Boy, and used a wire hook to catch the buckle beneath him, bringing the girth up to the near side. Shy Boy accepted the girthing without complaint; he did not buck or even jump. He never kicked at me, although he did strike out once or twice with his front feet when I moved too quickly for him.

After the surcingle was on, I decided that Shy Boy had had enough for one day. Both of us could use a night's rest. We had achieved join-up and were well on the way to putting a saddle and bridle on this wild horse.

By Monday afternoon Shy Boy seems content to be with Monty though he is still wary.

Pages 92–93: **With Caleb's help, Monty slips a rope over Shy Boy's neck.**

Shy Boy's first lead rope provokes a spirited response.

By this point I had been up for thirty-six hours, all of it in the saddle, and had done some of the most arduous riding of my career. I was exhausted. Strangely, though, it was hard to come down from that natural high. I have had a lifetime of athletic activity of one sort or another, but I have never enjoyed a second surge as I did that day. The major contributor, it seemed, was the moment of join-up. It gave me fuel that carried me for the rest of the day.

It was a major miracle that my back stayed together through this ordeal and I shall be eternally grateful for this bonus. Pat helped me take the membrane patches off my legs; the skin was in surprisingly good shape.

All the horses were fed and watered near camp, and the crew divided up night-watch duties.

I took some welcome nourishment and crashed. I slept well, but after five or six hours I was ready to go again. How could anyone sleep at a time like this?

Tuesday, Day 3, I had planned to be an easy, almost free day, so I could bond with Shy Boy on the ground. As a man standing alone on the ground, I was a different entity from the "man on horse" he had learned to trust.

The surcingle goes on to give the mustang a taste of the saddle and girth to follow.

Page 98: **Cathie and Caleb prepare the camp and settle in for a well-deserved rest.**

Page 99: **Horses on the picket, a warm fire going, and the setting sun in this glorious valley: a little bit of heaven.**

He accepted my rubbing and stroking him. I ran my hands all over him and down his legs. He followed me around and we worked with the surcingle again. It was like a day of rest for him, but he made good progress all the same. He had plenty of time to eat and drink and even lay down in the afternoon sun. We stood around together, walked together. Shy Boy was allowing me into his life and appeared to be under no pressure at all. This was a whole new world for him. We used Tuesday to deepen a relationship that I hoped would carry us into Wednesday—the day slated for him to accept his first saddle, bridle, and rider.

At daybreak on Wednesday, I rode George about a mile and a half to the west and down into a canyon we had discovered. From the high desert above, the canyon one hundred and fifty feet below was totally invisible. Only when you rode right to the rim did the spectacular sight below begin to unfold. The canyon was about three hundred feet wide for most of its half-mile length, with a stream meandering through.

From the banks of the stream arose huge California sycamores, some of them probably two hundred years old, with trunks six to eight feet in diameter. Bare of leaves, the sycamores' white bark and twisted limbs have inspired artists

Page 100: **Shy Boy, Dually, and Monty in a quiet moment at sunset.**

Page 101: **Monty and Caleb, youth and experience, both find comfort around the campfire as the temperature dips below freezing.**

Monty reaches out to stroke Shy Boy's head as a reward.

for centuries. This is the kind of place that Frederick Reming-ton, Charles Russell, or Ansel Adams might also have chosen, but for a different purpose all together—for art's sake.

We chose the place for its beauty, but for practical reasons too. The rocks here had been ground into sand by the water's action over thousands of years, and the stream had spread lay-ers of the sand over much of the canyon floor. Better still, the canyon floor was flat.

It would be safe for Shy Boy's feet and for the rider, too, should he be unfortunate enough to get bucked off.

We rode down an S-shaped trail to the canyon floor. Caleb was on Tari, a 16-hand blood bay. On George, I led Shy Boy beside me. The camera crew and some of the ranch hands used four-wheel-drive vehicles to ferry all the equipment into place.

Pat Russell and Scott Silvera had the riding equipment, a snaffle bit and bridle, my surcingle, a small exercise saddle, sad-dle pads, and a stock saddle. We were ready.

The surcingle was no trouble; we had been through that the evening before. Next I asked Shy Boy to accept the small exercise saddle, and he said, "No way." It was just too much leather to suit him. But after a little coaxing, he allowed me to put it on.

The stock saddle, on the other hand, struck Shy Boy as an insurmountable obstacle to the progress of our relationship. I

Tuesday morning and a long day ahead. Monty is hoping Shy Boy will accept saddle and bridle.

Page 106: **Preparing for Shy Boy's first saddle, Monty puts on a saddle pad.**

Page 107: **Monty approaches Shy Boy with the first saddle.**

spent half an hour persuading him we could be trusted. But join-up had clearly paved the way: once that first link is forged, the rest falls into place—if you're reasonable. Shy Boy accepted the stock saddle off and on, several times and from both sides. He accepted the bridle as well quite readily; by this time he was really looking for a friend.

Next came the biggest step of all: I introduced him to Scott Silvera, who would be the first person to ride him. Shy Boy was suspicious of the new face and reared away from Scott. We took time out for him and Shy Boy to become acquainted; the mustang smelled Scott's clothes, felt the rubbing on his neck and on the bony parts of his forehead.

Scott then slowly and carefully put a toe in the stirrup. Shy Boy jumped back and struck out with his front foot.

We all had to keep calm at this point. There should be a complete lack of urgency in any situation like this. Horses need patient handling. Act like you've only got fifteen minutes, it'll take all day; act like you've got all day, it might take fifteen minutes.

After taking more time for making acquaintances, Scott once more put his foot in the stirrup. Shy Boy accepted him. Scott lifted himself up very carefully, and swung his leg across. His bottom touched the saddle, light as a feather.

The desire to see this happen was buried deep, and watching it finally take place gave me immense joy. I felt like I had been given a great gift.

Placing the small first saddle on the mustang's back.

Allowing Shy Boy the feel of the saddle
and rewarding him for accepting it.

Pages 112–13: Monty approaching Shy Boy
with a western saddle.

I yelled out in triumph when Scott rode him off. Within five minutes, Shy Boy was a relaxed horse. We rode around the canyon, I on George and Scott on Shy Boy, for about half an hour to school Shy Boy and let him get used to the idea of being ridden while we were on this soft terrain. He did not put a foot wrong.

Shy Boy carried his rider as if he had been doing it half his life. The evidence was in his gait. A horse puts either the left or right foot forward first, and he may change that "lead" on his own or if asked by his rider. Shy Boy was naturally changing leads, a clear sign he was relaxed. Sometimes horses carrying a rider for the first time show signs of stiffness—psychological or physical—and they will not change leads. Shy Boy was very much focused on his rider, looking back at him often, but he was otherwise calm.

Next we intended to ride back to the Russell ranch house. I was on George, Caleb on his own horse, and Scott on Shy Boy. Three abreast, we cantered up to the house without a care in the world. I was virtually speechless with excitement.

Clearly, others had the same response. Dr. Miller was waiting there, along with Carol Childerly and the entire BBC film

Page 114: **Monty preparing to cinch up the saddle as Caleb looks on.**

Page 115: **Shy Boy seems to be saying that this cinch really tickles.**

Monty places the first bit into Shy Boy's mouth.

Page 118: **Later on, a moment of relaxation.**

Page 119: **With Caleb, Monty leads Shy Boy fully tacked.**

crew, who had been watching these events right from the start. We were greeted by a good-sized group of people, all of them clapping and cheering.

Some of our closest friends had come as well, optimistic that the adventure would end happily; John and JoAnn Jones and Brian and Cheryl Russell were all standing beside Pat and Laurel and the others as we rode to the ranch house.

We looked after our horses and then adjourned for a proper celebration with a barbecue, music, and storytelling as the sun set in the western sky. The BBC crew even managed to get on film Dutch Wilson and his pal from the Maverick Saloon in Santa Ynez.

"Well," said Dutch, "they can put a man on the moon, I guess he can bring back a mustang."

Later Shy Boy went quite easily into a trailer, where he joined Dually, The Cadet, and Big Red Fox for the trip home. From Flag Is Up Farms, Shy Boy was taken to live at Ron Ralls's place, in Buellton. Ron is a former student of mine who has his own training establishment just down the road from me.

This would be Shy Boy's home for the next year. I would see little of my own home, and almost nothing of Shy Boy in the year to come. The obligations of touring to demonstrate

Page 120: **The wild horse ready at last for his first rider.**

Page 121: **Monty, Caleb, and Scott Silvera during the first attempt to mount Shy Boy. "Are you kidding?" the mustang says. "Not a chance."**

Photograph taken during the first minute of riding.

Congratulations all around: Scott on Shy Boy,
Monty on Dually.

Page 126: Shy Boy, Dually, and Monty
enjoying the California landscape near
Flag Is Up Farms.

Page 127: Shy Boy, Dually, and Monty up in
the mountains in spring with a blooming
yucca plant in the foreground.

join-up and to promote *The Man Who Listens to Horses* took me to Europe and all over North America.

During all that time I thought about Shy Boy a lot. How was his training coming along? Was he happy? Would he remember the old cowboy who followed him around that night in the valley?

Another burning question came up wherever I traveled. The BBC documentary on Shy Boy had by this time been widely disseminated and people wanted to know: Now that Shy Boy has come in from the wild, is he happier now? If given the chance to be free again, would he take it and run?

It seemed to be a question I would have to answer.

Page 128: **Monty, Dually, and Shy Boy in the high country.**

Page 129: **Dutch and Freddy at the ranch house: old skeptics become believers.**

On the morning after, two friends reflect on victory over a major challenge.

Victory!—Shy Boy leads us home.

4

AGAINST CRUELTY TO HUMANS AND HORSES

Watching this slip of a girl in that round pen was a moving experience—there wasn't a heart that wasn't moved.

I told myself a long time ago that I would go to the ends of the earth to show what is possible between humans and horses. I would tell the world that the gentle way is the better way. And were you to catch a glimpse of my schedule and the time I spend on the road, you would know how much I have taken that assignment to heart.

For most of 1998, I was away from Flag Is Up Farms. I counted only fourteen days at home.

Demonstrations of join-up take place in arenas and riding

Monty getting acquainted in the round pen with a raw, unschooled horse.

establishments, wherever there is interest in seeing a gentle alternative to the harsh treatment of horses. I do in a round pen what I did with Shy Boy in the Cuyama Valley. Typically, I start two raw horses and explain to the audience what to look for.

I take an untrained horse and normally within seven to ten minutes have him joined up with me; he will follow me wherever I go in the round pen, his nose to my shoulder. Within twenty minutes, he will be standing quietly, with no restraint, while I ask him to accept the saddle, then bit and bridle. Within half an hour, there will be someone riding that horse. No harsh words, no whip or spurs.

This process, which I call "starting" rather than "breaking," would have taken my father far longer and would have inflicted a great deal of pain.

In dealing with horses, my whole approach has been to go in the opposite direction of cruelty. That's because my character has been shaped by what I have seen and experienced. Watching the traditional methods used by my father and others—forcing horses into submission and binding their legs—had a profound impact, as did witnessing my father's extreme violence to other people and the physical abuse I suffered at his hands.

Perhaps understandably, my demonstrations often become a flashpoint for those in the audience who have themselves suffered physical abuse. Some individuals faint when they see

Using the horse's own language, Monty adopts an aggressive stance to push the horse away and into flight around the ring.

what I do with a horse in the ring and talk about the futility of cruelty. Certain people experience a catharsis and feel compelled to tell me their stories.

Domestic abuse is more prevalent than I ever imagined. If, say, two thousand people come to a demonstration, up to half that number will approach me during one of the breaks—perhaps to have a book or video signed, or maybe to pass on a greeting. Typically, in the lineup will be some twenty to thirty individuals who have experienced physical abuse in their own families. It's not easy but they do talk about it.

There must be the same number again, and probably the worst cases, who avoid talking about abuse at all costs, and for good reason. Talking about it can break families apart. I know that from my own experience. Publication of my memoirs did not, to say the least, sit well with some members of my family. Admitting abuse threatens the family lineage and reputation.

Not to talk about these things is to disguise them or, worse, to condone them. Some young people recover from physical abuse, but for many the damage is irreversible, just as it can be for animals.

My experience on the road, starting horses before audiences, has left me convinced of the inextricable links between cruelty to animals and cruelty to humans. Saying no to the former inevitably raises hard questions about the latter.

This herd animal, longing for a friend, soon learns that the alternative to all this work of fleeing perhaps lies with the man in the ring who seems to speak the horse's body language.

I do what I can to address the violence. My events typically benefit charitable organizations in some way. I am happy to report that in 1998 alone such events raised $655,000 dollars for causes that will improve the lives of horses and horse people.

Proof that brutality works no better with horses than it does with humans came during a tour of Britain in 1996. Our first stop was in Essex, where I met an extraordinary girl.

A woman approached and said she was escorting my biggest fan, whereupon she reached down and lifted a child onto her hip. "This is Samantha," she said, and both showed me a photograph of Samantha's horse, Bess.

"She might be a fan," I replied, "but she's not a very big fan." I thought she was maybe eight or nine years of age. Samantha, I later learned, was twelve.

"She's read your book three times," confirmed her mother.

Samantha was several inches shy of five feet, with very slight features. Though small, she seemed healthy, with rosy freckled cheeks. Owing to my eyes (which, remember, see only shades of gray) I can't say what color her eyes were, but I imagined they were blue or green. What I most recalled later was their sparkle. She had beautiful white teeth and you got to see them a lot because she smiled constantly.

I was impressed by her bright, intelligent expression and her serious attitude. She promised she was going to learn every-

The moment of join-up. The horse and Monty have struck a deal. Now the horse is ready and willing to listen.

thing she could about horses. She said she would let me know how it was going, then asked if she might give me a hug and have her picture taken.

This direct emotional response touched me and I knew I would always remember her, whether or not we ever saw one another again.

Samantha, and a great many others I meet in my travels, frequently ask a basic question: "Can anyone do what you do?" More particularly, the question is "Could I do what you do?" In short, people want to know if they might follow my methods and achieve join-up themselves.

My answer is always "Absolutely." I might have been the first to happen upon the language of Equus, but if a person is determined and has the confidence and sufficient ability to recognize what to do and when, then he or she can achieve join-up. I have shown many competent horse people my methods and had them up and running in no time.

There are some practical considerations. This method is, first of all, not for those who are phobically frightened of horses. One should be relatively sound physically before attempting this technique. It is not something you can do from a wheelchair or crutches or with weak eyesight: you need to see the animal clearly to execute the language.

But the more competent students are at handling themselves around horses, the better their chances for success. Get

Adjusting the first saddle. This is not a major concern for the horse, because he has experienced nothing but gentleness from the man in the ring.

instruction from a professional and, above all, use safe practices when handling a horse—by these methods or any other.

Samantha, as promised, sent me a note about a year later, along with a homemade videotape. I could see Samantha in a building with a horse that quite plainly was responding to her. After looking at the tape several times, it became clear that Samantha had achieved join-up with her horse, Bess.

At this point, I was only eight days away from a demonstration to be held in London's Docklands Arena, which has a capacity of about five thousand people. It was also the venue closest to where Samantha lived.

I called her and invited her to attend. Could she possibly summon the courage, I asked her, to tell the audience what she had accomplished? I also invited her to bring Bess, her horse, to show everyone.

All who came will remember Samantha, and her story. When she was about three months old, she suffered from a severe case of flu that weakened her terribly; thereafter she spent much of her time in the hospital. She sometimes had fits, endured incredible pain, and suffered respiratory arrest. By the time she was five years old, she weighed about twenty-eight pounds. Samantha lived on the edge.

Eventually someone recommended a certain allergy specialist. Samantha's mother had grave doubts but took her to him in a last-ditch attempt to save Samantha's life. This little girl

Still, the saddle is a foreign object and the horse tries to buck it off but soon settles, especially since Monty remains resolutely calm throughout.

was put through a severe regimen of withdrawal from a wide range of foods and kept alive on the blandest diet possible before new foods were added, one or two at a time. By the third week certain allergenic foods, which had been weakening her system, had been identified; she was much better and, soon after, was allowed home. Samantha's parents finally had their daughter back.

They also had a promise to keep. During the crisis of a partial respiratory arrest, Samantha's desperate mother had tried to boost her daughter's fortunes by asking about her hopes and dreams. "Is there something you really want when you get better?" she asked. "A horse," replied her daughter without hesitation.

Samantha's mother was horrified. They knew nothing about horses. But a promise is a promise. Once Samantha reached the age of twelve and was off the rigorous diet and pills, her mother set about looking for a horse.

This is not an easy task, even for an expert. Anyone who has ever bought a horse has a story to tell about the experience. This purchase had quite a twist to it. Her mother read about a horse for sale and made an appointment to see her. The owners led the horse around and she seemed quiet and easy to manage. Impressed, Samantha's mother made two more appointments to view the horse, a tall 16-hand mare with Thoroughbred looks. Each time the vendors insisted on making an appointment, as opposed to Samantha's mother showing up unan-

First bridle. One more step in a process that invariably introduces the horse to his first rider in less than half an hour.

nounced to view the horse—and the last time she went, they had the horse tacked up. Everything seemed ideal. Samantha's mother then took her daughter to see the horse, again by appointment. The girl, of course, was overjoyed.

But when the mare arrived at her new stable, the unloading caused an uproar. She reared, pawed, and threw herself to the floor in an attempt to rid herself of her handlers. Samantha and her parents were assured it would just take a few days for the horse to settle in.

That did not happen. This horse had no intention of cooperating with anyone. She would rear and kick out and she refused to be ridden. Samantha's parents knew then that the mare must have been drugged before every visit. The horse was clearly vicious, and that was never so apparent as the time she turned on Samantha and bit off the top of her fingers, which had to be stitched back on. The mark of Samantha's character and courage, and her faith in that horse, is that despite all this she persisted.

Samantha's parents finally called in professional help and sent the horse off to be schooled. The mare managed to fling off an experienced horsewoman but eventually returned more tractable, though still unridable. It was at this time that Samantha saw the first BBC documentary describing my join-up methods.

She started badgering her parents about this Monty Roberts fellow. Finally, after some investigation, they realized that I was

Longlining the horse with thirty-foot reins so the horse begins to understand something about stopping and turning before the rider gets on.

not a pop star—as her mother first presumed—but a horseman. She purchased my book.

After studying it and, later, the video, and completely on her own, Samantha eventually gained enough confidence to attempt join-up with her horse. To her delight and amazement, it worked. Her horse, named Raichia, turned out to be a Russian-Arabian with papers, so Samantha's parents won in the end.

They later sold Raichia—as a quiet, lovable horse who had developed into a reliable mount.

Samantha's next horse, Bess, had also been difficult when first purchased. However, she did have a lovely temperament and they wanted to stick with her. At first when they put a saddle on her, Bess bucked, carried on, and bolted. The parents again sent their horse to a professional trainer and thereby discovered Bess's problem: the horse had been tied down, climbed all over while on the ground, and abused. The sight of a whip sent her into hysterics.

Bess was not as severe a case as Raichia, and join-up clearly worked with her too. Samantha now has Bess going beautifully in the show jumping ring and is winning awards.

Watching this little slip of a girl in that round pen in the London arena was a moving experience. It was gratifying to see her talking the language of Equus and using it so successfully with Bess. Samantha was the star of that show and deservedly so. She overshadowed me and everyone else. There wasn't a

The rider, too, is careful to mind his equine manners. Here he introduces himself to the young horse before getting on.

heart that wasn't moved. In that arena you could feel the glow of admiration for her.

She had saved a horse from an untimely end in a slaughter-house and spared herself further harm—and she had done it without ever raising a hand or even her voice.

Samantha's story was about getting past the abuse that someone had inflicted on her horses. But everywhere in the world, I know, are children subjected to the same raised fists and blind discipline. The man I would call the Memphis cow-boy offered a grim reminder of that fact.

We were on tour in Tennessee, one of those stops where we were to do demonstrations on successive nights. I was halfway through the first evening, signing books and videos. In the line that had formed in front of me I noticed a huge man. He had a red face and while he wasn't much over forty, the sun had made its marks, leaving fissures on skin the texture of leather.

He wore a meticulously creased cowboy hat that had obvi-ously seen a few seasons of hard work. The hat held its shape, attesting to its quality and the fact that its owner was a pro and proud of it. Years of dressing movie actors in an attempt to make them look authentic had taught me how difficult it is for a nonprofessional to achieve "the look."

When he was a dozen or so people from reaching me, he stepped slightly out of line. His shirt, blue jeans, and boots

First rider. No bucking, no spurs, not a hand or voice raised.

were all correct for a top cowboy. He faced squarely on me just before returning to the line but not before I caught a glimpse of a trophy buckle with a few years on it and obviously earned in rodeo competition.

I spent many years as a professional cowboy, and it's not hard to spot another. They don't often get in a line for anything, these cowboys, least of all for someone's autograph. The curiosity in me heightened as he worked his way to the front. Maybe he'll say he rodeoed with me at some point, I thought. This sort of thing happens often. Maybe he'll remind me of an old friendship we once had. Then again, he may not want my autograph at all; perhaps he wants me to know that he's a professional "horse breaker" and that my work is a load of crap—though likely he would use another word. This too has happened.

"I want to thank you, Mr. Roberts," he began. "I've really learned something tonight. I don't have a book. All I have is this little piece of paper that I'd like you to sign, but that's not really important. I would have stood in this line for two more hours just to shake your hand."

With that he put an enormous hand across the top of my signing table. My own hands are big and strong, yet mine seemed small in his. His demeanor, though, encouraged me to put my hand into that freckled vise and to let him squeeze out his appreciation for what he had seen. I will never forget how he looked me in the eye and said, "You taught me something tonight, something I don't believe any other man could. I'm going to get your book."

As he walked away, I realized that he had been alone in that line. That struck me as unusual. Being alone is okay when you're working on the ranch, but when you come to town, you

just don't line up with a bunch of city folk without a few ranch hands for company. I must have signed for another hour or so, but I could not get that cowboy out of my mind.

The following night's event started and ended routinely. I was once more on the signing stand when I realized that the cowboy was back, standing just off to my left in an open, uncrowded area. Well over six feet four inches tall and built even stronger than I remembered, he stood about two yards from four little girls, aged eight to eleven, I guessed. All had long red curls down their backs, and all wore smart dresses. The cowboy stood with a book in one hand and a look on his face much like the night before, stern and studious. When he approached the stand he came to the side, not the front—a cowboy's way of coming to the family door, not the front door. I winced a bit at the thought of another handshake, but I braced for it because I was indeed happy to see he had returned.

"I read your book since I saw you last night," he said, "the whole thing. I never slept. Don't read too fast, either. Never read another book clear through in my whole life. You've done good things, Mr. Roberts. You're bringing good things to horses. I'll never treat horses the same again. My father was brutal to them, too, just like yours. But what I want to know is, is it too late for those four little girls over there?" With that he began to sob. I put an arm around his hulking frame and escorted him to a horse trailer nearby to get him away from people who were staring.

The little girls came slowly along, staying eight or ten yards between us. They waited, huddled together, while their father related stories of his brutality—to his horses, his daughters, and his wife. His wife, he said, couldn't come along tonight; she

feared too much what he was going through and felt safer at home. He admitted to being arrested on two separate occasions, when friends had turned him in, but his wife denied being beaten and got the charges dropped. He told me that she had covered for him on another occasion when he was questioned about reports that the girls had been severely beaten. With his arms around me, he pleaded for help.

We later made some calls and put him in touch with an agency in his community. He is now under their care and counseling and, at last report, is doing quite well.

Before leaving me that night, he held the four little girls in two massive arms and pleaded for their forgiveness. He promised never to mistreat them again and asked if they would please tell their mother what he had said.

If this one evening were all the payment I ever got for a lifetime of work, it would be enough.

5

THE CASE OF
BLUSHING ET

He came at me with ears back and teeth bared. I believe he could
have killed me any time he chose.

Shy Boy, like the many thousands of horses I have started in my life, had a gentle introduction to saddle and rider. But do gentle starts ensure gentle ends?

"It's all very well giving your horse this benevolent start in life," I am often told, "but how does that translate, later on, in the adult horse?" Dually, my world-class western competition horse, is the answer to that question. A chunky registered quarter horse, he was a pigeon-toed, bow-legged character that would give any buyer pause.

His name comes from a type of pickup truck with powerful rear axle and wheels to take extra load, a hint of the strength and power in his rear quarters. Quick and agile, he works with

hardly a touch on the reins. He is my number one friend and colleague in the showring.

But Dually is full of the devil. A horse with a lot of energy, he thinks he owns the farm, the trees, fences, and all. He's arrogant and full of personality. When I am with him he wants all my attention. Should I talk to someone or fail to keep him front and center, he will nip at me or push me with his nose to remind me that this is his time with me.

Dually accords only one horse a position above him, and that's Shy Boy. Possibly Dually understands how important the mustang is to me. Shy Boy may not know it, but he has been granted a great privilege from King Dually.

What Dually cannot do in western-style show events cannot be done. He is an intelligent and willing horse: he can cut a cow from a herd and hold it at bay, separate from the herd—whatever twists and turns the bovine might try. He will do this of his own accord, without a sign from me. Dually knows exactly what to do and wants to do it for himself. It's all I can do to hang on.

Dually is a true example of what gentle training methods can achieve. If you can use your skills as a trainer to open a door that a horse wants to go through, then you have a horse as a willing partner instead of your unwilling subject.

The word "unwilling" barely begins to describe Blushing ET, perhaps the greatest challenge I have faced as a trainer.

Monty and Dually one misty morning at Flag Is Up Farms.

Dually, it turned out, would have a role to play in helping this troubled horse see the light.

The story began when Sleeping Giant, a television production company in Toronto, decided in mid-1997 to produce a documentary involving an extremely troubled horse. The horse was to be chosen from a list of Thoroughbreds known to be paranoid about starting gates. These are the twelve to twenty-four stalls set out in a line to ensure a fair start at racetracks. The goal was to film that one horse's journey: from fear, to acceptance, and back into racing. I was to be the trainer in the film; the search then began for a worthy horse.

Crawford Hall, the farm manager at Flag Is Up Farms, contacted trainers, starting-gate crew, and jockeys. Among them was Sean McCarthy, a former farm employee and the student who had spent more time than anyone working with me on rehabilitating horses terrified of starting gates. Sean happened to be with me in England when I was called to Henry Cecil's "yard" in Newmarket to deal with North Country, a troubled horse that eventually came right and won races.

Responses to Crawford's queries began to pour in from all over the United States—proof of the severity of this problem in racing. It was Sean, though, who came up with Blushing ET. The horse had been placed in training with the racehorse trainer Janine Sahadi at the Santa Anita, California, racetrack. Both trainer and track are among the best in North America. Janine had been warned that Blushing ET had a problem with

Dually and Monty share a rare quiet moment at the farm.

the starting gate, but little did she know of its severity. The first trip to the gate resulted in disaster: a broken hand for the handler, Zane Baze, and a horse loose on the track.

Blushing ET was a two-and-a-half-year-old sired by Blushing John, and that line tends to be intense—as many great racehorses are. The horse had been sent to California from Florida, where alarming stories surfaced: aggressive beatings reportedly ended with Blushing ET lying flat on the ground and refusing to get up for ten to fifteen minutes at a time. The horse had become extremely aggressive and had injured handlers while being forced to enter the starting gate, and even while lying on the ground thirty yards from the gate. All agreed: Here was a hopeless case.

I accepted Blushing ET as my challenge and even entered the project with confidence, especially after I achieved normal join-up with him in the round pen. But I had woefully underestimated the determination of this horse never to allow the starting gate back into his life. I have had a lot of experience with horses petrified of starting gates. Prominent in my mind are Lomitas, North Country, and the horse that for seven years ranked as my greatest challenge in this category, Prince of Darkness.

Blushing ET was in a league of his own.

I remember being deeply embarrassed that bringing Prince of Darkness to a point where he could race had taken me eight days. Blushing ET would take me eighty.

Dually is a supremely gifted horse who thinks he owns the farm.

Where Prince of Darkness had a huge problem with the starting gate and a lesser one with human beings, Blushing ET had a violent phobia about the starting gate and a deep-seated hatred for humans. I believe the hatred was justified and born of abject brutality. I should have been more sensitive to these differences.

Blushing ET was transported to Flag Is Up Farms in November 1997, and I began my work supremely confident that I would send him back to Ms. Sahadi with time to spare for an opening date at Santa Anita on December 26. By the end of the first day's session, I realized I was dealing with a new dimension in troubled horses.

Picture this: a beautiful chestnut Thoroughbred with correct legs and gazellelike movements. He was tall at 16.1 hands and still growing. Horses are measured in hands: one hand is four inches, and the measurement goes from the ground to the withers, the high point of the back. He had yet to come into full maturity but he was already strong and athletic. He was a nervous, angry horse.

Blushing ET began very early to dive at me with teeth bared; all because I tried to put a protective blanket on him. The first two or three days were devoted simply to securing the blanket and getting him to enter an enclosed area I had constructed on each side of the gate itself. Going into the gate by moving forward was out of the question.

Neither was I was able to back him into the gate. Experience has taught me that the horse unwilling to back on the lead

Blushing ET explodes the second anything touches his hocks. He was the greatest challenge Monty ever faced as a trainer.

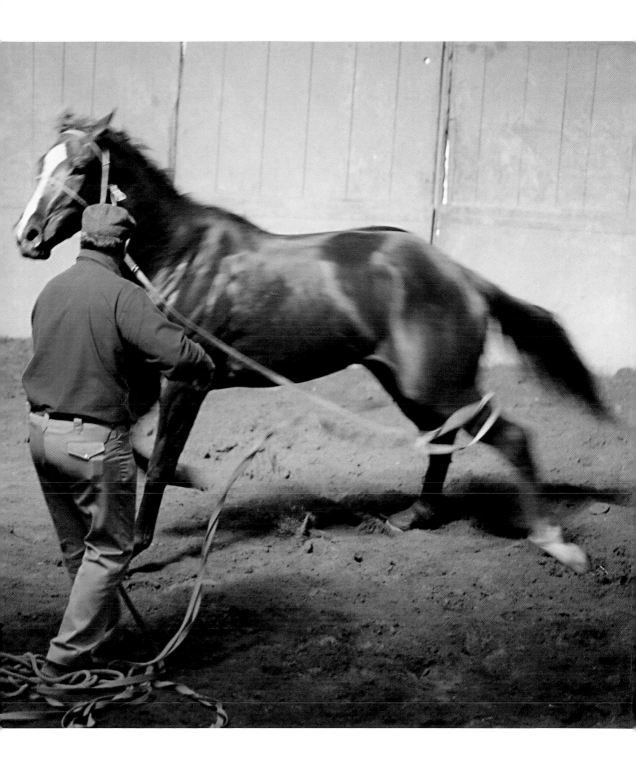

will nearly always fail to move forward on the lead. Blushing ET would not back a step.

To keep his aerobic fitness intact, we gave him a good canter each day as well as a ride around the farm from time to time. We hoped to freshen his mind and encourage him to see human beings in a better light. Exercise rider Felipe Castro worked with him every day, but the problem at the gate showed little overall improvement.

Blushing ET gave me the feeling that there just might be a piece of the puzzle missing. He taught all of us much about the effect of trauma. I could go only so far with him before hitting a brick wall, but I could not put my finger on the cause of it.

He was kind and peaceful in the stable; the farmhands could groom and bandage him and generally manage all his stable requirements. Even washing him and running water down his legs did not bother him. But put anything foreign around his hocks—the joint in the hind leg that points backward—and he would explode, ears back and thrashing. This was a serious problem: a groom's brush touching his hocks, a blanket strap, a longline—any of those things were like setting a match to a stick of dynamite.

I had a strong suspicion that all of this had to do with the jockey's foot rails, which travel along each side of the individual starting stalls and meet the horse squarely at the flank and the stifle, extremely sensitive areas. The stifle is the joint at the highest portion of the hind leg, equivalent to our knee.

Blushing ET seems to trust Monty early on but an epic eighty-day struggle lies ahead.

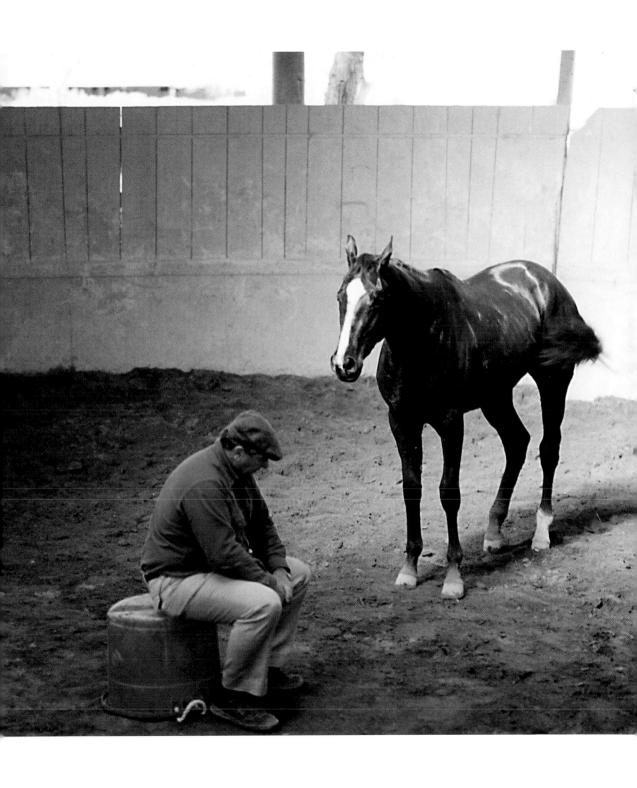

Blushing ET was so beset by those side rails that it seemed I would get nowhere until he was convinced they would not harm him. I was astonished that using the blanket plus hours of work had not had more dramatic effect. The blanket is a device of mine that offers horses frightened of starting gates a sense of security and protection. It's a little like what a picador's horse wears in the bull ring, and it simply falls away when the race starts.

Even putting Blushing ET on the longlines was impossible. Blushing ET wanted none of it. If you dropped the line over his hip and along his side, he went beserk. He would kick out angrily with his back feet and strike the ground with his front feet—as though he would do anything to escape those lines. It seemed certain he would hurt himself, or me. Sometimes his aggression sent a clear message: "I'm going to come and get you."

He would march around me, a line on each side, then attack with such force I was unable to pull him off me. I believe he could have killed me anytime he chose. I confess that during those moments I had absolutely no control.

One evening I was watching a football game on television when I realized I barely knew who was playing. My mind was fixed on Blushing ET. I decided to tackle this problem. After mustering some assistance, I drove to the round pen. It was already dark when I got there and I put the lights on. Two members of the farm staff, Caroline Baldock and Faith Grey,

The moment of join-up between Blushing ET and Monty. The starting gate, though, still held terrors for the horse.

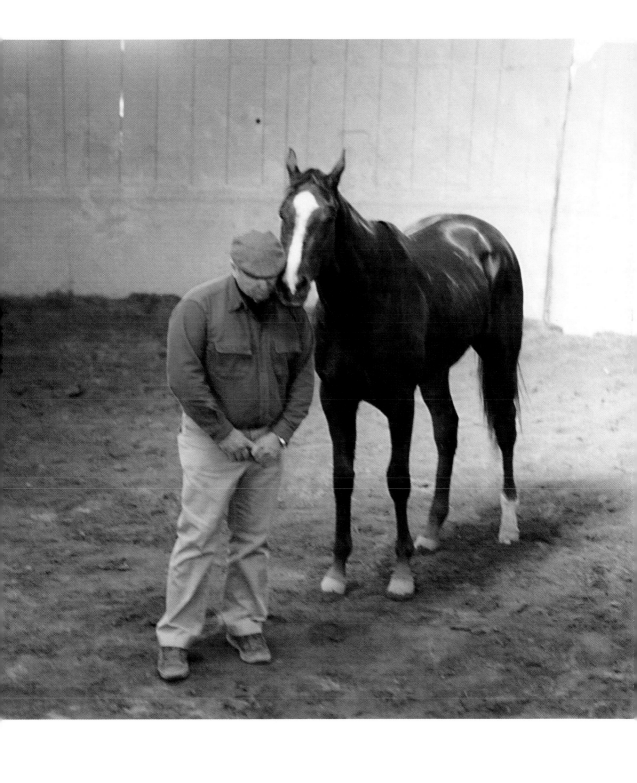

were both with me. Still, I felt very much alone when I entered the round pen.

When we arrived at the stables, I first went in and saddled Dually, checked the reins back to the saddle horn and left him standing in his stall. To this day, I don't really know why I did that. Perhaps subconsciously I knew I might need Dually's help. In all my years of resolving starting-gate dilemmas, involving dozens of other horses, it had never occurred to me to enlist the help of a second horse.

I then saddled Blushing ET but left his stable blanket on. This was to lessen the effect of the line near his hip and stifle. I felt confident that with the blanket in place I could encourage him to accept the line, and later I could remove the blanket. Once more I gave his phobias far too little respect.

In the round pen my plan was shattered almost immediately. He was just as violent as he had been without the stable blanket. I did everything I knew to get him to settle and accept the lines.

I was now in deep trouble. As the minutes went by, he became more angry. Despite my calm and deliberate manner, he would dart at me with ears back and teeth bared, coming at me in full frontal attack. I tried to double him around, pulling the off line to get his head away from me. It didn't work. He was too strong. I began to fear for my safety.

Caroline and Faith jumped down from the viewing stand and opened the gate so I could escape to safety. I walked around outside, trying to get my own heartbeat down. The more I thought about it, the more I realized his deeply felt fear of anything along his sides. The abuse he had received as a result of trainers' attempts to cope with his natural aversion to the

starting gate had only compounded matters. The harder they whipped him, the more he blamed the rails and, later, the human as well.

I walked up the ramp to the viewing deck, where I could see him dragging the longlines around the pen. It dawned on me that by using strips of cloth tied to the stirrups, I could perhaps get him used to objects along his sides and near his hocks. These cloth strips would tear away under the pressure of his violent kicking, unlike the driving lines, which could entangle his feet and lead to serious injury.

I went into the stable and got some saddle towels and girth covers; the latter are like three-foot-long tube-socks, and they're designed to prevent skin infections from being transferred from one horse to another. The cloths, I thought, would at least let Blushing ET get used to something flapping at his side. The longlines would come later.

Blushing ET, meanwhile, had relaxed enough in the round pen to let me remove the longlines and tie the cloths to the stirrups. He kicked them and fought them and then finally began to settle a bit. It seemed he would grow accustomed to them. I gradually lengthened the cloth strips until they touched the ground and came back around his hind legs as he moved.

Again he kicked at them, but I felt I was now in a position to try the longlines again. A violent crack, the sound of his hooves slamming the side of the pen, told me I was wrong. This time he had almost nailed me with those deadly feet.

He then charged and nearly got me in his teeth. Once more I slipped out the gate, and those deadly front hooves were less than an inch from making contact. Faith and Caroline were really worried by now and very much wanted me to quit.

I paced up and down outside the pen racking my brain. All of a sudden the light came on. What if I mounted Dually and rode him into the round pen? Would Blushing ET accept Dually? I would be up off the ground and, I hoped, in less danger.

I rode Dually into the round pen and gathered up the long-lines. I put one over Blushing ET's back, attached the other directly to the bit, and began to drive him—without putting the lines over his hips. By *drive,* I mean directing the horse, so he moves and turns according to pressure from the thirty-foot-long reins.

This was more or less acceptable. I could actually put my reins over the saddle horn and guide Dually with just my legs and voice. After a few minutes, it seemed it was going well enough. Again I put the longline over Blushing ET's hips to continue driving him. As soon as the line went over the hip, he kicked, bit, and struck out with his feet, and twice made strong attacks on Dually. I was able to jump Dually forward and avert Blushing ET's charge, throwing my arms up to get him to stop.

It would be a stretch to say I had the upper hand while in the saddle, but at least I was in a neutral position and not a wholly vulnerable one. I took the longline off the near side and left the one on the off side. I began to ride in a big circle so that as I crossed behind ET, I could bring the line near his hock. When it touched his hock, he would kick fiercely.

I would allow this kicking until there was the slightest pause and then immediately I would take the line off his hock, riding in a circle around his front to bring the line close to the other hock. I would repeat the process on the opposite side, quickly rewarding any improvement by once more riding away. Soon I was able to achieve a pattern: I could hold the longline

against his hock and the kicking, by degrees, subsided. I would then ride in a big circle, sometimes at a trot, and let the line touch the other hock. I would then repeat the process.

It was amazing to watch Blushing ET learn to communicate with me. I could see enlightenment come over him. When he stopped kicking, I would take the line off his hock. After a period of time, I could let the line touch his hock; he wouldn't kick and I'd ride away. The instant that happened, I was on a course to succeed.

If all learning is zero to ten, then the most important part of learning is zero to one.

After about a half hour of this work, I was able to put both lines on, run them through the stirrups of the saddle, and drive him around the pen while up on Dually. The more steps I took, the better he got. Every once in a while, he would lightly revert and I would go back to square one. These periods of regression rapidly became shorter, however, and it seemed I was making great progress.

Before the evening was over, I was able to drive him around the pen without incident. There was an intermittent bout with failure when I decided he was going well enough that I could step off Dually and drive Blushing ET from the ground. This was a mistake and immediately he reverted to aggressive mode.

That was the moment I realized the depth of this horse's hatred for men on the ground. I subsequently noticed that anytime a man came near him during loading in the starting stalls, he would worry. If anyone walked behind him, you could see him get ready to fight. I suppose he is still that way to a degree and I doubt he will ever forget what other humans have done to him— beaten him in an attempt to make him enter the thing he feared

most, the starting gate. But that night, the hours between six and nine-thirty P.M. marked a course change for ET and me. After that, using longlines and with me standing behind him on the ground, I could drive him through the starting stalls.

I could drive him walking in circles, right through the starting gate and eventually without a surge of panic. It was now a matter of using his own language, going step by step through each of the procedures, and he soon began to behave more like a normal horse.

I am deeply grateful to Blushing ET for all he taught me. This horse reminded me that we learn when we are challenged. My task was to come up with a nonviolent way of communicating my intent while respecting this horse's rights.

It may seem strange that I would commend a horse for his patience when it looked for all the world like he was trying to annihilate me. The fact is, he did not annihilate me. And given what the horse had been through and the beatings he took, the miracle was that he did not kill four or five people before ever coming to me. Yes, Blushing ET was patient: he was an effective teacher who waited countless hours for me to learn. He taught me to understand a level of fear—but tenacity, too, and spirit— that I had never seen in a horse before. Until I could convince him that the longlines weren't going to kill him, there was nothing I could do to get him near a starting gate. Nothing.

Felipe Castro cruises through the starting gate on Blushing ET. Before this could happen the horse sent Monty back to the drawing board countless times.

More than any other horse, he taught me to back up. If I ever thought I was a genius or some world-famous horse trainer, Blushing ET reminded me that I was in kindergarten.

Blushing ET provided me with the information I needed to come up with a new device—I call it "the hallway"—that could help horses like him. It works on two principles: repetition is the heart and soul of learning, and herd animals find comfort from moving in circles. Were you to look down on this device from a height, here is what you would see: a miniature racetrack, an oval with two little straightaways, and two little turns—the whole thing perhaps one hundred feet long from start to finish. The track is defined by seven-foot-high panels, and the horse enters the oval through one such detachable panel.

The riderless horse is led, typically counterclockwise, around the track. At one point on the oval the inner stall of a three-stall starting gate—typically used to introduce young horses to actual starting gates—has been placed; the outer two stalls sit outside "the hallway." It's important that the gate be left open: this whole exercise is about teaching the horse gate manners without applying any pressure. The horse's first impression, therefore, is of a passageway; later comes the notion of a gate.

The horse might spend many hours on that track, until the routine is established. You go forward by degrees. Add a saddle, then use the longlines. Later add a rider, then stop at the gate briefly before proceeding. Finally, one day horse and rider enter

Blushing ET back on track and winning races.

the gate, which closes behind them. Eventually, it closes in front and behind.

If the horse industry comes to understand "the hallway," it may become a feature of racetracks all over the world. Several tracks in the United States are interested in testing it.

Sadly, there are still horses everywhere trying to tell human beings that they are being treated unfairly, but the message is not getting through. For so many years I myself failed to understand why horses saw fit to fight starting gates. My only solace is that nobody else did either. I do hope that trainers and people who make decisions about horses in racing will listen to the lessons of Blushing ET.

Many trainers still don't get it. The rails down the sides of starting gate stalls are not user-friendly to horses. These gates are noisy, unnatural, forbidding. They're big, tall, skinny, made of cold metal, and the people around them—like everyone at the track—are full of adrenaline. Horses sense it immediately. I'm amazed that so many horses willingly enter these starting gates.

Blushing ET, meanwhile, is back on track. Knowledgeable people in the American racing industry are calling this nothing short of a miracle. He won two races consecutively, one a maiden race at Hollywood Park on July 12, 1998, and another, an allowance race, at Del Mar on the turf with a purse of $55,200 on August 26, 1998. He is healthy and sound and perhaps has more victories in store for his owners.

Blushing ET may not like the starting gate or the handlers that jockey him into position before a race, but the abject terror and fear that this tall chestnut felt for that cold metal seem, mercifully, to have subsided.

6

SHY BOY'S NEW LIFE

The farrier stormed into Crawford's office. "What is this wild thing you've brought me?"

If the Easter weekend of 1997 marked a turning point in my life; it was surely the same for Shy Boy. The little bay mustang had entered the world of the domestic horse—a far cry from all that he had known in the Nevada foothills.

Like many captured mustangs, the first thing he did in his new life off the range was to grow taller and put on weight. Land left to the mustangs is typically rugged, arid, rocky: life there is often hard and the grazing meager, and the constant need to keep moving means mustangs tend to stay lean. When I first saw him that day in Paso Robles, Shy Boy was about 13 hands, 2 inches. He now stands about 14 hands, 1 inch, and weighs about 975 pounds, about 100 pounds more than he did on the range.

As I write this, almost two years have passed since I gentled

him, and he still possesses some of the instincts of the wild horse. It's hard for him to put his head all the way down into a feed bucket, for example. With his eyes covered, he can't spot enemies. He still worries about predators, even in the safety of his stall. Eating hay from a net over his head is also something he does warily.

And he takes water whenever he can, as his mustang heritage has taught him: one day I was telling students at the farm that mustangs always drink water when they see it. Shy Boy was close by in a paddock. There had been a downpour that day and the mustang, as if on cue, was stopping at every puddle, sticking his nose in and sucking up a mouthful of water.

But what has emerged during all this time is his striking personality. Widespread airing by PBS in North America in 1997 and 1998 of the BBC documentary on Shy Boy has made the little mustang something of a celebrity. People have come to our farm from all over the continent and many parts of Europe just to see this horse.

And clearly he loves the attention. Pat has observed him in the paddock when crowds come around. "He preens very proudly," she says, "and arches himself as if he were posing. And he loves kids. He goes right to them and tries to lick and nuzzle them."

Shy Boy knows what he wants, and sometimes what he wants is the company of children—as we found out in the summer of 1998. The road to our training barn runs alongside a

Tara Twissleman with Shy Boy as brother Caleb looks on. The mustang has developed a great affection for children.

paddock he sometimes enjoys, and this day one of our staff noticed a car stopped on the road. To the great amusement of his rider, a horse had stuck his head and neck deep into the car. The horse in question was Shy Boy.

The rider, it turns out, was Felipe Castro. Along with Jason Davis, he has spent a lot of time riding the mustang and looking after him. Felipe had noticed his brother in the passing car and stopped exercising Shy Boy to say hello. The horse was curious to see who was in the car, and since the window was open, he put his head right in the driver's side. He was hoping for a face rub, his usual greeting.

But it was not Felipe's brother he was really interested in; it was the little girl in the passenger seat. He probed deeper into the car. Once the girl had rubbed his forehead and Shy Boy was satisfied that everyone in the car had greeted him in the proper manner, he carefully stepped back and away.

Shy Boy, like many good horses, has developed over the years an acute sense of his own importance. Because so many people fuss over him, he seems to think he should be the center of attention. He likes to be noticed and can get quite cross if he is bypassed by people.

One day my farm manager, Crawford Hall, was showing some people around the training facility when they passed Shy Boy's stall. Suddenly there was a racket: the mustang raked his teeth across the stall screen and bit one of the boards. "Oops," said Crawford. "We made a mistake. We passed Shy Boy's stall without saying hello." The mustang got the rub he thought was his due.

He looks to me like a contented and well-adjusted horse, but naturally there have been some ups and downs. The first test of his fabled mustang toughness came early. He had barely settled in to his new digs when a virus hit the little horse and almost leveled him.

Within weeks of getting Shy Boy to the farm, I had made a decision. The mustang would continue his education as a working western horse at the ranch of Ron and Billie Jo Ralls, just three miles down the road from our place. I had good reason for choosing them.

Billie Jo came to work for me in the mid-1980s and stayed on for eight years. It became immediately apparent that she brought to her job on the breeding side of the farm's activities a significant amount of experience. Within months she was making a lot of the decisions for that operation and soon became manager. Billie Jo could care for the medical needs of the mares as well as or better than most veterinarians.

She can dress up and look fabulous, but were you to go out to their ranch you might find her running the tractor or mucking out stalls. She is a woman of immense energy and if you shook her hand you would be struck by the strength of her handshake.

Ron Ralls is a lean, sinewy six-footer with a ruddy complexion, and, like Billie Jo, he's extremely fit. He and Billie Jo married while both were working for us. They later started up their own operation, breeding and training horses of the highest caliber for western competition. Pat and I have two outstanding prospects with them now, Peppy San Nic and Captain Nice.

The Ralls ranch is set well back in the foothills, and the remarkable thing about it is the mesa, or tableland, that over-

looks the ranch. They've carved a racetrack up there, where the natural footing is all beautiful loamy sand. The infield is an oval of some fifteen acres, with facilities for holding cattle, cutting, and roping. This would be Shy Boy's new school ground.

I knew that his training and care at the Ralls ranch would be the best. At first, Ron was less than keen on the idea. When I broached it, he looked at me out of the corner of his eye as if to say, "I'm doing fine right now. I've got some nice horses that might very well win major championships. Do I really need to be riding a mustang?" When I explained how important Shy Boy was to me, he agreed to take on the task of training him.

After ten days, Ron had a fairly good understanding of Shy Boy. Ron never forgot Shy Boy's wild origins and took this into consideration in all his schooling. Ron knew that Shy Boy could spook and act out, as any wild horse would. Sometimes, young ranch hands—assigned to ride the mustang up the trail to the mesa where Ron was to work him—were shocked by his little explosions. A piece of plastic blowing in the wind, say, would send him tearing back down the hill. But the mustang never did that with Ron, who became his anchor.

Billie Jo, for her part, also got to know the horse during those first ten days in the stable, and she, too, became aware of just how wild he was. Her keen eye may well have saved Shy Boy's life.

On the evening of the eleventh day, Billie Jo entered his stall and saw immediately that his demeanor had changed significantly. He didn't jump when she opened the door; he didn't

Monty working with Shy Boy in the round pen.

stiffen his muscles or go to the back of the stall and snort through his nose—all the precautionary measures he had been taking for every one of the past ten days. He was no longer fearful and protective of his environment. Something was wrong.

Billie Jo stopped in her tracks and said aloud, "Shy Boy, are you OK?"

She stood there for a moment just inside the stall door, looking at him. He walked slowly up to her and put his head near her. She stroked his head and repeated, "Are you OK?" It hit her that the horse seemed warm, extraordinarily warm. She left the stall, went to the tack room, and got a thermometer. With a bit of coaxing, Shy Boy allowed her to insert a rectal thermometer. Reading it a few minutes later, Billie Jo immediately concluded that the device was broken. It read 107.2 degrees. Normal temperature in a horse is 100 degrees, slightly higher than what's normal in humans, 98.6.

Billie Jo had cared for a lot of sick horses in her time, but she had never taken a reading like that. Nor have I. Billie Jo plucked another thermometer from the first-aid kit in their travel trailer, and took a second reading. Same result.

Fear shot through her. Mustangs tend not to get the flu virus; on the other hand, they have neither immunity to the virus nor the vaccinations that would protect them. Resisting panic, she called Dr. Van Snow—who happens to be Ron's brother and a vet in whom we all have a great deal of confidence. She then called Crawford and advised him that Shy Boy was in trouble. Finally, she took a third reading using yet another thermometer. There was no doubt about the reading.

Billie Jo wet a towel and stroked Shy Boy's head, and with

that, he appeared dramatically changed. He seemed to be saying he needed a friend, for he knew he was in jeopardy. What amazed her was how he would take a few sips of water and even munch on a bit of hay every few minutes. Under these same conditions, a domestic horse would almost never do that. Hit with a red-hot flu virus, a domestic horse will immediately go off feed and water. It's a huge problem because it leads to dehydration.

Shy Boy seemed to know that eating and drinking were good for him. And unlike domestic horses, who typically want to lie down as a response to sickness, Shy Boy seemed determined to remain on his feet.

When Dr. Snow arrived, his first concern was that the mustang lacked the advantages of the immunization programs that domestic horses have. This fact limited the medications he could use on Shy Boy. Those he would normally use might have worsened the fever, with possibly fatal results. After some deliberation, he decided on phenylbutazone ("bute") and sulfamethoxanzole trimethoprim (SMZ). It was hoped that the bute, like aspirin an anti-inflammatory, would knock down the temperature. The SMZ was intended to boost his fluid levels.

All they could do now was wait. Billie Jo set up a cot outside Shy Boy's stall and slept there.

I learned of all this when I was touring in England, eight thousand miles away. I was devastated and felt quite helpless. Yet I had absolute faith that the Ralls would do everything they could. No matter the outcome, there was no blame to place.

"Leave no stone unturned," I told Billie Jo on the telephone.

During that first night, just after midnight, Shy Boy's temperature began to subside, and by daylight it was down to 102.8. But he was not out of the woods yet.

Billie Jo told me that in the second and third day of the illness, Shy Boy's tenacity, his strength, his innate survival instincts took over and saved his life. He still refused to lie down and seemed to stiffen his resolve to eat and drink.

There was some discussion about putting him on intravenous fluids, which would require inserting a tube into his jugular vein and placing a bottle dripping on a stand next to him. Not the kind of thing a mustang would tolerate. But his persistence in eating and drinking removed the need for that option; he continued to recover. Billie Jo stayed with him for three full days and nights and kept him on the bute and SMZs for ten days.

During that time, Dr. Snow ran tests to see if they could reduce the medication. But each time they did so, the mustang's temperature spiked upward again. It became clear that Shy Boy was harboring a strong and potentially lethal virus.

By the end of the ten days, Billie Jo had the mustang's temperature stabilized and the medications reduced. He spent another thirty days recovering at Flag Is Up Farms, and when he was his old self, he returned to Ralls ranch to continue his training.

In a typical day, Shy Boy would be saddled and ridden. At first the workouts focused on basic things like walking, trotting, and cantering. Later his daily routine would include figure eights, with flying changes, stops, and spins—all in preparation for his work with cattle.

Ron began to work cattle on Shy Boy in the second month

he had him and that was the primary thrust of his daily routine. Each training session would last about an hour, after which Shy Boy would get a bath and some time in an outside paddock.

But it was clear from our conversation that Ron had fallen in love with the mustang. Ron is a man of few words, and when I asked him why he liked Shy Boy so much, all he could say was "He is just a neat little horse." I went to Billie Jo and asked her to find out more about why Ron liked him. She came back to me the next day and said, "He just thinks he's a neat little horse."

I think I can translate what the phrase means to someone like Ron Ralls. I believe the key word is "willing." Back that word up with "generous." You might add in "cooperative." Ron's thinking likely goes like this: compared to his superstar, world-class horses, Shy Boy is small, and lacks strength and speed, but every day he tries to overcome these shortcomings by giving you everything he has.

Ron probably thinks that if you put that kind of courage and willingness into those highly bred superstars, you could improve any one of them. To be deprived, for his first three years of life, of a high-powered nutritional and medical program puts him so far behind the stars in the making that he can never catch up. But that doesn't mean he won't try.

A good coach frequently develops a soft spot for a student who is never going to make the championship ranks but whose work ethic, desire to please, and generosity set the youngster apart. That's why they have awards for "most improved" or "most dedicated." Shy Boy falls into this category.

One thing Shy Boy had to overcome is his natural inclination to protect his feet. Mustangs guard their legs and feet—those parts of the anatomy that allow them to flee—with extreme zeal. My experience with mustangs has taught me that you can touch them virtually anywhere you want before you can pick up a leg or foot.

Shy Boy, no surprise, was wary of anyone touching his feet. During his time at the Ralls ranch, he learned about bathing and grooming. He learned it was all right to have someone clean his stall while he was in it. He learned about leading other horses while he was being ridden and many other facets of life at a training facility. There was no need to have him shod, of course, because his mustang heritage had already given him strong, sturdy hooves. But Ron did slowly condition Shy Boy to accept having his feet cleaned and trimmed. The trimming would keep the hooves level and counter uneven wear.

"It took a bit of time when I first did it," said Ron, "but after that it was fine. No problem."

During a time when Shy Boy was back at Flag Is Up Farms, Crawford had a look at Shy Boy's feet and decided it was time for a trim. He put the mustang on a list and thought no more about it. A few hours later, David Bowen, then our farrier, burst into his office. "What the heck have you given me to trim? This horse Shy Boy just told me in no uncertain terms that I was not allowed to pick up a foot on him. I tried all four corners. I don't have a shot."

"Let's have a look," said Crawford. He made his way to the area where the farriers work. As they approached Shy Boy, his eyes like saucers, Crawford asked David how long he thought it

might take to trim the mustang's feet. No longer than fifteen minutes, David replied.

Crawford reminded David of a precept I hold dear: "If you try to do something with a horse as if you have fifteen minutes, it might take you all day. Let's go at it like we have all day," said Crawford. "There's a fair chance we'll come closer to getting it done in fifteen minutes."

Moving more slowly now, and showing Shy Boy that he was someone to be trusted, David got the job done in short order. The same scenario unfolded two months later with another farrier. "What is this wild thing you've brought me?" he complained. "Hasn't he ever seen a pair of nippers before?" Again, more time was spent on the task, and Shy Boy learned to trust the new farrier.

Mustangs are incredible horses. But you need to understand them and respect their need to trust. Fifty-five million years of history have taught them to be cautious. I can easily load Shy Boy in a trailer, but he may balk if I send someone he doesn't know to do it. Asking him to walk over unfamiliar objects or through tight places will produce the same result: he reads his human partner and responds accordingly.

Shy Boy had something else that distinguished him from the other horses on our farm. He had worms. Long worms, stomach worms, blood worms, intestinal worms.

The mustang takes his water where he can get it; oftentimes it will be stagnant, muddy, and laden with parasite eggs. The result is that most mustangs are infested with internal parasites, and young mustangs, especially, are vulnerable to infes-

tation. Yearling mustangs typically show severe signs of intestinal parasites, but as the horses age they build up an immunity. If I could wave a magic wand and rid every mustang in North America of parasites, they would all put on a few hundred pounds and grow a hand taller.

Domestic horses, on the other hand, generally get their water from clean sources. In addition, horses raised in stables and paddocks are wormed regularly.

Anyone who adopts a mustang has to be cautious about building up a worming program. The medications themselves are a poison, of sorts, or they wouldn't kill the worms. Dr. Snow set up a precise worming program for Shy Boy and the other two mustangs we had adopted.

The worming medicine was started slowly, and Shy Boy passed what seemed to be huge amounts of internal parasites. Giving him the normal dosage for a domestic horse right from the start might have been disastrous. Shy Boy remained healthy and as his system rejected the parasites, which would have sapped him of energy and strength, you could see him take on a glow of health and a feeling of exhilaration that he likely never would have experienced as a wild horse. He's now on the normal worming procedure, and he bucks and plays, expressing himself in a far more gregarious manner than was first apparent.

One day I took him from his stall and led him toward a field; my intention was to turn him out so he could get some grass and exercise and sun. As I unlatched the gate, Shy Boy

Two horses nuzzling.

began to squeal in anticipation. He jumped a few feet off the ground, straight in the air, then landed before doing it again. He looked like a horse on a pogo stick and I had to laugh at his energetic clowning. When I led him into the field and released him there he ran around for about twenty minutes.

He looked to be celebrating the mere fact of being alive.

The other two mustangs, I told Crawford, are at the farm solely to teach my students. And what a job these two horses have done.

The youngest one, a dark bay named Mustang Sally (he's a gelding, but never mind), could well have been my horse of choice for the film. He has a prettier head than Shy Boy's, with a long, elegant neck and a body like Shy Boy's. Mustang Sally is just as bright and flighty and responsive as Shy Boy is, but a year younger—that and his small size took him out of the running for the film.

He's now taller than Shy Boy by almost an inch. A good mover, very elegant for a mustang, and very pretty. Perky too, with his ears always up.

The other horse, a lighter bay than Shy Boy and a year older, is called Franco. He's a dictator. He wasn't as quick to become cooperative; he still tells you how much of a mustang he is. Franco is now bigger than both Mustang Sally and Shy Boy, with

Shy Boy and Monty in the round pen. The mustang loves attention but he also loves to romp in the paddock.

feathers on his chin and legs. He tucks in that head like a sea horse and he demands, and gets, permission for everything.

If I could manage it, I would have thirty mustangs at Flag Is Up Farms, just to teach students. For a student to spend a week with Franco, say, is like spending six months with a domestic horse. With Franco, you better put on a halter or brush him the proper way. He charges you a price whenever you're wrong. The domestic horse is thicker, less inclined to show displeasure at your mistakes.

That makes mustangs wonderful teachers. These wild horses are so raw, so green, so pure.

7

SHY BOY'S RETURN

Hundreds of people asked this one question: Is Shy Boy better off as a ranch horse, or would he rather be wild? It dawned on me that the one to make that choice was Shy Boy himself.

This book was inspired by Shy Boy.

Gentling him virtually in the wild had unfolded almost without a hitch, though not without anxiety and challenges along the way. The BBC documentary that chronicled the adventure was subsequently shown in North America and has led to countless inquiries.

Literally hundreds of people asked me this one question: "Do you think Shy Boy is better off as a ranch horse, or would he rather be wild?"

Pages 198–99: Shy Boy enjoying a wild gallop on a spring morning.

I don't know, I would reply. I can't read his thoughts. I can observe a steady, contented horse who is happy in his work and affectionate in his dealings with those people responsible for his care. But I can't judge whether or not he longs for the open range.

This much I did know. After spending almost all of 1998 on the road, I saw Shy Boy in December of that year. And I wondered, would he know me after all that time? It seems he did. In a matter of seconds he knew who I was and considered me the center of his world. In a large crowd he would look right past everyone, even the person leading him, and head toward me.

There is one way to tell if a horse enjoys doing something: give the horse the opportunity to do it and see if he does the activity voluntarily. Using the same logic, I often tell students, "You say your horse loves to jump. Well, put him in a field full of jumps and if he chooses to run around and jump them all by himself, I'll believe you." It's not such an outrageous idea; some horses have been known to do just that.

In the case of Shy Boy, many people who knew his story wanted to know his preference. Given the chance, which would he choose? The life of a domesticated horse, or that of a mustang? It dawned on me that the one to make that choice was Shy Boy himself. We would take Shy Boy back to the wild and there let him decide.

Monty and Dually looking out across the high desert.

The Twissleman family was due to round up three hundred and fifty head of cattle early in March 1998, almost a year after I joined up with Shy Boy on the Russell ranch and neighboring lands in the high California desert. The Twissleman ranch nearby is enormous, even by California standards. My guess is it's forty or fifty miles from one end to the other. The ranch is located in the same high desert we had been the previous year, when I gentled Shy Boy.

I suggested that Caleb Twissleman ride Shy Boy on the roundup, and that at an opportune moment, he would leave Shy Boy in the company of the same free-range herd that he had originally been taken from. We would give him the choice: he could stay with the free ones or come back to us.

To record that moment—just as we had captured on film the gentling of Shy Boy—I hired a film crew, four-wheel-drive trucks, a plane for a few overhead shots, and everything else I would need to document Shy Boy's decision. Dr. Robert Miller, who had served as the veterinarian and animal behaviorist on the BBC documentary, agreed to be with us, start to finish. He would observe and comment on the psychological aspects of the study and oversee the safety and treatment of the animals involved.

All agreed that this would be a valid postscript to the Shy Boy story. We went into the planning stage. I decided I would ride The Cadet for the high country part of the roundup. Caleb

**Monty and Dually in a casual canter through
sagebrush country.**

and I chose the western part of the spread because that was the area where the free-ranging herd were most often spotted.

On Monday morning, March 9, we loaded our horses on the trailer. Shy Boy, The Cadet, and Dually were fit and ready to go. Pat and I had also asked Caroline Baldock to join us. I had met her in England; she's been a friend for eight years and for several years my research assistant. Her task was to oversee the outline for the project, ensuring that all the salient parts of the story were captured on film.

We headed north on Highway 101 through the coastal towns of Santa Barbara County and San Luis Obispo County, turning inland at Santa Margarita. We proceeded through some of the most beautiful mountains our state has to offer, toward the high desert where we had been eleven months earlier. Winter rains had turned the country a deep Irish green and the wildflowers were just beginning to unfurl across the south-facing slopes.

We drove to a point more than three thousand feet above sea level before leveling off on the Carissa Plains. The whole area was covered by a green carpet of winter grass less than an inch high, with barely a tree in sight. Soda Lake was visible thirty miles away, silvery and flat as a sheet of glass cut to fit the dip in the land.

The San Andreas fault runs right through the middle of this vast expanse. We were to camp right over it, on the north margin of the plain. As Pat and I stood there, observing the crew

Shy Boy and Caleb ready to round up cattle on the sprawling Twissleman Ranch.

Cathie Twissleman, after a long winter,
eager to move cattle in the spring.

arranging the trucks and trailers, I was suddenly struck by the grandeur of this plain framed by huge mountains. The breeze carried a scent of sage and clover. The silence was palpable.

Pretty soon they had a fire going in the camp. The welcoming smell of burning oak permeated the air.

Just then a high-pitched wail cut the silence and a chorus of howls answered. It was like a thousand dogs were all answering the same question. The pack of coyotes could not have been more than a quarter of a mile away. It was frightening to those in the group not familiar with coyotes because the sound curdles your blood; but to those of us who know how harmless they really are, it was just a reminder that we were back in the wilderness.

Later, during the night, one of the crew reported to me that Shy Boy would occasionally stiffen and prick his ears. Thirty seconds later, she herself would hear the coyotes. The mustang's radar could sense them long before the human ear.

It was cold that night, just as it had been eleven months earlier when it seemed I would freeze before ever touching Shy Boy.

At sunup Rowly Twissleman and his wife, Cathie, and son, Caleb, came driving into camp. They had two huge trailers, one for horses and one for equipment. We prepared for the day in very short order and Rowly split us up into various teams. The job at hand was to gather cattle and it was time to go. Only when that was done would we put the question to Shy Boy.

Ranch hand Zane Davis on Arnie before the roundup.

Pages 210–11: **Caleb on Shy Boy, with Monty behind on Dually, moving cattle along the trail.**

As the sun came up and began to warm us, I felt a wave of contentment wash over me. The beauty of that landscape, the old familiar joy of camping out, and the thought of a day moving cattle—all that worked a little magic in me. With Caleb on Shy Boy and me on The Cadet, we headed for the northeast corner of this vast acreage.

Caleb was born and raised here and knows every nook and corner. We went up through a canyon and over a ridge and found ourselves moving along the face of a steep mountainside through sparse chemise brush and sage. A jackrabbit jumped out and nearly stopped my heart. The Cadet took flight and I fought to regain my seat. Caleb laughed at the sight. Shy Boy hardly took notice. He had seen plenty of jackrabbits in his time.

Soon after reaching the very top of the ranch land, we began to assemble a sizable number of cattle. We started working them downward near the east margin of the property. We had about sixty head by then, but I could see many more out in front of us.

We were now probably two thousand feet higher than the corrals, and through the binoculars I could see the other teams driving cattle toward the pens. It seemed everything was running smoothly.

Caleb and I pressed on, picking up more cattle along the way. At one point we dipped into a canyon with a significant

Caleb and Shy Boy bringing in cattle. On these drives you aim to move at a cow's walking pace, but you must always be ready to scoot.

flow of water at the bottom of it. The odor of rotten eggs greeted us and grew worse as we descended. A sulfur spring, Caleb explained.

It was time for the horses to drink, but The Cadet wouldn't go near the water. Shy Boy, however, was happy to take a drink and Caleb allowed him a few mouthfuls. Shy Boy was educated in the ways of nature and eleven months of domestication had not made him forget that this water would not hurt him. Besides, nature also had taught him it might be a long time before he saw water again.

As we were beginning our descent to the holding corrals, I noticed about ten head off to one side of our herd. I suggested to Caleb that he go on with the main group while I galloped off to gather these extras. As I approached the cattle, I noticed Shy Boy's free-range herd in the brush beyond them. I turned the cows toward Caleb's main group and did nothing to disturb the horses.

We drove the cattle down to the holding corrals, where two hundred head or so were waiting for us. From our vantage a few hundred feet above the pens, the beauty was stunning. Stretching out in front of us were about thirty miles of the Carissa Plains. The floor of the desert was awash with wild flowers. Almost four hundred cattle were strung out below us and dotting the landscape with their colors: the red and white of the Herefords, the jet black of the Angus, the gray, blue, and buckskin of the Brahmans.

Monty on The Cadet.

SHY BOY

And then there was Shy Boy, cantering to the flank of the herd, jumping ditches to dodge the sage. We guided our cattle into the holding corrals, and I was pleased to see that Shy Boy was a valuable member of the team. He actually appeared quite proud of himself.

Once we had the cattle corralled, it was time to have some lunch. The atmosphere was festive around the picnic tables, with the music of Garth Brooks blasting from the cab of one of the vehicles. We took off the saddles, fed and watered the horses, and hosed down their legs. An hour or two of rest was welcomed by all, animal and human.

One of the ranch hands had reported that the free-range herd, Shy Boy's bunch, had begun to graze closer to the camp, probably intrigued by our horses (and the sound of Garth Brooks).

With lunch and a good rest behind us, it was time to cut out the designated yearlings from the herd. Now weaned, the young cattle would be taken to feedlots, where they would fatten on grain. I was on Dually. He was doing what he loved to do best—cutting cattle. He's happy to do it in competition, but he loves doing it for real.

Caleb on Shy Boy skirting around the outside. Note that Shy Boy has all four feet off the ground.

Page 218: **Monty and Dually cutting cattle as Shy Boy looks on to see how it's done.**

Page 219: **Pat delighted to see the two old pros working in the cutting pen.**

Caleb had remounted Shy Boy, and together they were watching Dually, the old pro, cutting out the young cattle. Caleb called out that he'd like to give Shy Boy a try and I agreed. He must have been watching Dually closely, as he seemed to get the hang of it in very short order. Shy Boy earned a round of applause from our group for preventing a steer from getting back to the herd.

By the time we had finished cutting out the yearlings, it was four in the afternoon and the shadows were getting long. A campfire was lit and the horses were settled down with a good flake of hay. Cattle not slated for the feedlot were released.

The sunset painted colors on the underside of the clouds and the air was filled with the sounds of cows calling and calves replying. As more than three hundred cattle went tramping through the sage, the air grew pungent with that aromatic mixture that I have long associated with riding in open country.

It was nearly five o'clock when Caleb and I sat down by the fire to discuss the business now at hand—the release of Shy Boy. The free-ranging horses, ranch hands reported, were just out of sight on a ridge north of camp. The wranglers took a few of their saddle horses toward the southeast to tempt the range herd to investigate. It worked perfectly and the herd soon appeared atop the ridge in sight of the camp. There were about twenty of them, silhouetted as the sun set, edging them in gold.

While Caleb looks for cattle across the canyon, Shy Boy eyes the photographer.

SHY BOY

Shy Boy was the only horse in camp to notice them. His ears were locked in their direction, his big black eyes focused on the herd. I said to Caleb, "This is what we came for; it's time to see what happens." Caleb's face wore a mixture of apprehension and sadness. "Give him a rub," I said, and he did.

The instant I took his halter off, Shy Boy whirled and in one graceful jump he was in a full gallop headed straight for the herd on the horizon. Fear shot through me. Shy Boy moved with absolute certainty.

It was nearly dark, yet we could see him clearly when he joined the horses on the ridge. I thought the herd would reject him, at least in the beginning. After all, for eleven months he had been on different feed and water. He had lived almost a hundred miles away from them; surely they had forgotten him. I was amazed to see just the opposite; they accepted him without a murmur, as if he had never been away. There were no challenges, no biting or kicking. The last thing we saw before darkness was Shy Boy out front, leading the herd over the ridge.

Shy Boy had gone straight to them, with alarming gusto. I had to work hard to convince Caleb that it wasn't over yet. Shy Boy had had nothing but kind treatment from us, I told Caleb. He had had a warm bed, good food, and fresh water every day of the eleven months we had him. I felt confident he had had a wonderful time with us, but would that be enough to swing the decision in our favor? I wasn't sure.

Everyone, Shy Boy included, has spotted the free-ranging herd on the horizon.

Tara, Caleb's eight-year-old sister, had no doubts the mustang would return. She's a freckly strawberry blonde, as outgoing and animated as her brother is quiet, and she's very comfortable around animals. "Don't worry," she told us. "Shy Boy's coming back. You'll see."

In my own brain, the question "What have I done?" kept squirreling around. We had given Shy Boy the option, confident in his return. But serious second-guessing now plagued me. There was a lot at stake. If the mustang chose to live out on the range, I would have interpreted that decision as a hard knock to principles I hold dear.

My work with horses is predicated on the idea that if you treat an animal well and feel something for that animal, the affection works both ways. I believe that if you establish a partnership with a horse, the horse will want to be with you.

In fact, there's a whole school of thought suggesting that in the very early days of man's working relationship with animals, it wasn't humans that domesticated animals but more the other way around. Wild dogs would follow human encampments, hoping for food. Horses may have done the same, seeking protection from predators and feeding on cultivated crops. There were clear advantages to being around humans.

Horses, then, if treated well, don't feel incarcerated. We had never forced Shy Boy to do anything.

But Shy Boy had left me in a terrible hurry. I didn't sleep all night. I walked around making excuses that there were things I

Dr. Robert Miller, veterinarian, watching over the entire project.

Monty and Caleb. Shy Boy is gone and all they can do now is wait and hope he returns.

Pages 228–29: Attempting to bring him back with a silent prayer, Monty wonders if he can live with Shy Boy's choice.

Shy Boy heads for home.

had to do, but found myself doing little but watching the skyline around the camp looking for a familiar silhouette.

Maybe, I thought, Shy Boy was out there remembering some of the difficulties he faced in the first three years of his life. Summer days when the temperature hit 100 degrees and there was no water. I am sure there were winter nights when the temperature dropped well below zero. At one time or another, he may have felt the deep fear brought on by the sight of predators, or a narrow escape.

Maybe he wasn't sleeping much either as he pondered his decision.

At five A.M. I wandered through the horses and the cattle in the pen. I continued to watch, but there was nothing. By six A.M. we could see the hillside clearly enough to know that Shy Boy was nowhere near us.

Meanwhile, Rowly Twissleman had a cattle drive to wrap up and a great many practical matters to attend to. He called out his orders: stoke up the fire and get the breakfast going. We had to drive all these yearlings back to ranch headquarters. Caleb, though, was dragging his feet.

We all ate breakfast as slowly as possible, hoping to delay long enough so Shy Boy could make an appearance. With the mustang gone, Caleb had, of course, lost his mount; compounding his sense of loss was the prospect of having to ride home in a truck. There was tension around that morning fire.

He's heading straight for Monty.

I saddled The Cadet and decided to brush out his tail. Strange the things you do when trying to kill time. Then I heard Tara call out, "Hey, look there. It's Shy Boy. He's come back!"

I looked up sharply and I could see movement in the sage-brush. There was a single horse, about three hundred yards to the north, on a hill above us. The rising sun cast a bright light across the ridge, and we could see clearly that he was moving in our direction. The horse trotted right to the center of a clearing in the sagebrush and stopped, like an actor pausing on a stage. I was once more struck by how handsome Shy Boy is.

Everybody in the camp was stunned into silence. Shy Boy remained motionless. It crossed my mind that even if he whirled and took off again, at least he had come back to say good-bye.

Shy Boy, with a deliberate movement, turned his head to look back in the direction of the herd. He held that position for a few seconds, then looked toward the camp. He lowered his head and began to walk.

Shy Boy was truly a free horse because he had a choice. Was he still making up his mind? The direction he took was oblique to us. And he moved slowly, deliberately, as if considering which way to turn. As he reached the edge of the clearing, his head lifted and he broke into a trot, coming squarely in our direction now. He moved in serpentine fashion through the brush.

Pat and Monty clearly delighted with Shy Boy's decision.

Some two hundred and fifty yards from the camp, he broke into a full gallop. There seemed to be a path through the sage and he followed it in zigzag fashion. In my mind I was repeating over and over, "Come on, Shy Boy, come on home." Caleb was standing directly behind me now.

Running at a full gallop straight toward us, Shy Boy gave a loud, clear whinny.

We did not move a muscle. I wore a big smile and stood happily in his path. He galloped full out and only at the last second did he apply the brakes and come to a halt a few yards in front of me so I could step forward and welcome him in.

Shy boy had chosen to come home.

Caleb's sister, Tara, relieved after Shy Boy's return.

With Shy Boy posing, Pat shows Monty how she plans to
sculpt the mustang's image.

Pat's relief sculpture of Shy Boy.

Pat's Shy Boy sculpture showing the mustang
in a gallop.

Monty's tours are often family affairs. Marty Roberts and his mother, Pat, at one of the events.

Monty's Family

Pat and I have been married since 1956. Pat, a successful artist, exhibits equine sculptures on tour. She is equally at ease at the mike describing to audiences how her subjects were conceived and telling amusing anecdotes about me.

Our son, Marty, practiced law for eleven years prior to taking over the day-to-day operation as CEO of Monty and Pat Roberts, Inc. Marty not only organizes the demonstration tour, he also acts as emcee at the events.

Debbie, our firstborn, is married to retired tennis pro turned investment counselor Tom Loucks. They have two sons, Matthew and Adam, budding young tennis stars in their own right.

Our daughter Laurel has the green thumb of the family and is responsible for the parklike atmosphere visitors enjoy at Flag Is Up Farms, our home.

"Yucca"

"Tree Sculpture"

Christopher Dydyk
Fine Art Photography
christopher@montyroberts.com
(805) 686-5385
www.dydyk.com

"Happy Sky"

"Double Exposure"

For More Information

My goal is to leave the world a better place, for horses and people, than I found it.

For further information regarding the companion video to the book titled *Shy Boy: The Horse That Came In from the Wild* and for clinics, educational videos, merchandise, and other information please call:

toll free	1-888-U2-MONTY
	1-888-826-6689
on-line	www.montyroberts.com
e-mail	admin@montyroberts.com

Thank you,
Monty Roberts